1,000,000 Books

are available to read at

Forgotten Books

www.ForgottenBooks.com

Read online
Download PDF
Purchase in print

ISBN 978-1-330-49403-5
PIBN 10069389

This book is a reproduction of an important historical work. Forgotten Books uses state-of-the-art technology to digitally reconstruct the work, preserving the original format whilst repairing imperfections present in the aged copy. In rare cases, an imperfection in the original, such as a blemish or missing page, may be replicated in our edition. We do, however, repair the vast majority of imperfections successfully; any imperfections that remain are intentionally left to preserve the state of such historical works.

Forgotten Books is a registered trademark of FB &c Ltd.
Copyright © 2018 FB &c Ltd.
FB &c Ltd, Dalton House, 60 Windsor Avenue, London, SW19 2RR.
Company number 08720141. Registered in England and Wales.

For support please visit www.forgottenbooks.com

1 MONTH OF FREE READING

at

www.ForgottenBooks.com

By purchasing this book you are eligible for one month membership to ForgottenBooks.com, giving you unlimited access to our entire collection of over 1,000,000 titles via our web site and mobile apps.

To claim your free month visit: www.forgottenbooks.com/free69389

* Offer is valid for 45 days from date of purchase. Terms and conditions apply.

English
Français
Deutsche
Italiano
Español
Português

www.forgottenbooks.com

Mythology Photography **Fiction** Fishing Christianity **Art** Cooking Essays Buddhism Freemasonry Medicine **Biology** Music **Ancient Egypt** Evolution Carpentry Physics Dance Geology **Mathematics** Fitness Shakespeare **Folklore** Yoga Marketing **Confidence** Immortality Biographies Poetry **Psychology** Witchcraft Electronics Chemistry History **Law** Accounting **Philosophy** Anthropology Alchemy Drama Quantum Mechanics Atheism Sexual Health **Ancient History Entrepreneurship** Languages Sport Paleontology Needlework Islam **Metaphysics** Investment Archaeology Parenting Statistics Criminology **Motivational**

THE LIFE

OF

GALILEO GALILEI,

WITH

ILLUSTRATIONS OF THE ADVANCEMENT

OF

EXPERIMENTAL PHILOSOPHY.

MDCCCXXX.

LONDON.

LIFE OF GALILEO:

WITH ILLUSTRATIONS OF THE ADVANCEMENT OF EXPERIMENTAL PHILOSOPHY.

Chapter I.

Introduction.

THE knowledge which we at present possess of the phenomena of nature and of their connection has not by any means been regularly progressive, as we might have expected, from the time when they first drew the attention of mankind. Without entering into the question touching the scientific acquirements of eastern nations at a remote period, it is certain that some among the early Greeks were in possession of several truths, however acquired, connected with the economy of the universe, which were afterwards suffered to fall into neglect and oblivion. But the philosophers of the old school appear in general to have confined themselves at the best to observations; very few traces remain of their having instituted *experiments*, properly so called. This putting of nature to the torture, as Bacon calls it, has occasioned the principal part of modern philosophical discoveries. The experimentalist may so order his examination of nature as to vary at pleasure the circumstances in which it is made, often to discard accidents which complicate the general appearances, and at once to bring any theory which he may form to a decisive test. The province of the mere observer is necessarily limited: the power of selection among the phenomena to be presented is in great measure denied to him, and he may consider himself fortunate if they are such as to lead him readily to a knowledge of the laws which they follow.

Perhaps to this imperfection of method it may be attributed that natural philosophy continued to be stationary, or even to decline, during a long series of ages, until little more than two centuries ago. Within this comparatively short period it has rapidly reached a degree of perfection so different from its former degraded state, that we can hardly institute any comparison between the two. Before that epoch, a few insulated facts, such as might first happen to be noticed, often inaccurately observed and always too hastily generalized, were found sufficient to excite the naturalist's lively imagination; and having once pleased his fancy with the supposed fitness of his artificial scheme, his perverted ingenuity was thenceforward employed in forcing the observed phenomena into an imaginary agreement with the result of his theory; instead of taking the more rational, and it should seem, the more obvious, method of correcting the theory by the result of his observations, and considering the one merely as the general and abbreviated expression of the other. But natural phenomena were not then valued on their own account, and for the proofs which they afford of a vast and beneficent design in the structure of the universe, so much as for the fertile topics which the favourite mode of viewing the subject supplied to the spirit of scholastic disputation: and it is a humiliating reflection that mankind never reasoned so ill as when they most professed to cultivate the art of reasoning. However specious the objects, and alluring the announcements of this art, the then prevailing manner of studying it curbed and corrupted all that is free and noble in the human mind. Innumerable fallacies lurked every where among the most generally received opinions, and crowds of dogmatic and self-sufficient pedants fully justified the lively definition, that "logic is the art of talking unintelligibly on things of which we are ignorant."*

The error which lay at the root of the philosophy of the middle ages was this:—from the belief that general laws and universal principles might be discovered, of which the natural phenomena were *effects*, it was thought that the proper order of study was, first to detect the general *cause*, and then to pursue it into its consequences; it was considered absurd to begin with the effect instead of the cause; whereas the real choice lay between proceeding from particular facts

---

* Ménage.

to general facts, or from general facts to particular facts; and it was under this misrepresentation of the real question that all the sophistry lurked. As soon as it is well understood that the general *cause* is no other than a single fact, common to a great number of phenomena, it is necessarily perceived that an accurate scrutiny of these latter must precede any safe reasoning with respect to the former. But at the time of which we are speaking, those who adopted this order of reasoning, and who began their inquiries by a minute and sedulous investigation of facts, were treated with disdain, as men who degraded the lofty name of philosophy by bestowing it upon mere mechanical operations. Among the earliest and noblest of these was Galileo.

It is common, especially in this country, to name Bacon as the founder of the present school of experimental philosophy; we speak of the Baconian or inductive method of reasoning as synonimous and convertible terms, and we are apt to overlook what Galileo had already done before Bacon's writings appeared. Certainly the Italian did not range over the circle of the sciences with the supreme and searching glance of the English philosopher, but we find in every part of his writings philosophical maxims which do not lose by comparison with those of Bacon; and Galileo deserves the additional praise, that he himself gave to the world a splendid practical illustration of the value of the principles which he constantly recommended. In support of this view of the comparative deserts of these two celebrated men, we are able to adduce the authority of Hume, who will be readily admitted as a competent judge of philosophical merit, where his prejudices cannot bias his decision. Discussing the character of Bacon, he says, " If we consider the variety of talents displayed by this man, as a public speaker, a man of business, a wit, a courtier, a companion, an author, a philosopher, he is justly the object of great admiration. If we consider him merely as an author and philosopher, the light in which we view him at present, though very estimable, he was yet inferior to his contemporary Galileo, perhaps even to Kepler. Bacon pointed out at a distance the road to true philosophy: Galileo both pointed it out to others, and made himself considerable advances in it. The Englishman was ignorant of geometry: the Florentine revived that science, excelled in it, and was the first that applied it, together with experiment, to natural philosophy. The former rejected with the most positive disdain the system of Copernicus: the latter fortified it with new proofs derived both from reason and the senses."*

If we compare them from another point of view, not so much in respect of their intrinsic merit, as of the influence which each exercised on the philosophy of his age, Galileo's superior talent or better fortune, in arresting the attention of his contemporaries, seems indisputable. The fate of the two writers is directly opposed the one to the other; Bacon's works seem to be most studied and appreciated when his readers have come to their perusal, imbued with knowledge and a philosophical spirit, which, however, they have attained independently of his assistance. The proud appeal to posterity which he uttered in his will, " For my name and memory, I leave it to men's charitable speeches, and to foreign nations, and the next ages," of itself indicates a consciousness of the fact that his contemporary countrymen were but slightly affected by his philosophical precepts. But Galileo's personal exertions changed the general character of philosophy in Italy: at the time of his death, his immediate pupils had obtained possession of the most celebrated universities, and were busily engaged in practising and enforcing the lessons which he had taught them; nor was it then easy to find there a single student of natural philosophy who did not readily ascribe the formation of his principles to the direct or remote influence of Galileo's example. Unlike Bacon's, his reputation, and the value of his writings, were higher among his contemporaries than they have since become. This judgment perhaps awards the highest intellectual prize to him whose disregarded services rise in estimation with the advance of knowledge; but the praise due to superior usefulness belongs to him who succeeded in training round him a school of imitators, and thereby enabled his imitators to surpass himself.

The biography of men who have devoted themselves to philosophical pursuits seldom affords so various and striking a succession of incidents as that

---

* Hume's England, James I.

of a soldier or statesman. The life of a man who is shut up during the greater part of his time in his study or laboratory supplies but scanty materials for personal details; and the lapse of time rapidly removes from us the opportunities of preserving such peculiarities as might have been worth recording. An account of it will therefore consist chiefly in a review of his works and opinions, and of the influence which he and they have exercised over his own and succeeding ages. Viewed in this light, few lives can be considered more interesting than that of Galileo; and if we compare the state in which he found, with that in which he left, the study of nature, we shall feel how justly an enthusiastic panegyric pronounced upon the age immediately following him may be transferred to this earlier period. "This is the age wherein all men's minds are in a kind of fermentation, and the spirit of wisdom and learning begins to mount and free itself from those drossie and terrene impediments wherewith it has been so long clogged, and from the insipid phlegm and *caput mortuum* of useless notions in which it hath endured so violent and long a fixation. This is the age wherein, methinks, philosophy comes in with a spring tide, and the peripatetics may as well hope to stop the current of the tide, or, with Xerxes, to fetter the ocean, as hinder the overflowing of free philosophy. Methinks I see how all the old rubbish must be thrown away, and the rotten buildings be overthrown and carried away, with so powerful an inundation. These are the days that must lay a new foundation of a more magnificent philosophy, never to be overthrown, that will empirically and sensibly canvass the phenomena of nature, deducing the causes of things from such originals in nature as we observe are producible by art, and the infallible demonstration of mechanics: and certainly this is the way, and no other, to build a true and permanent philosophy."*

Chapter II.

Galileo's Birth—Family—Education—Observation of the Pendulum—Pulsilogies — Hydrostatical Balance—Lecturer at Pisa.

GALILEO GALILEI was born at Pisa, on the 15th day of February, 1564, of a noble and ancient Florentine family, which, in the middle of the fourteenth century, adopted this surname instead of Bonajuti, under which several of their ancestors filled distinguished offices in the Florentine state. Some misapprehension has occasionally existed, in consequence of the identity of his proper name with that of his family; his most correct appellation would perhaps be Galileo de' Galilei, but the surname usually occurs as we have written it. He is most commonly spoken of by his Christian name, agreeably to the Italian custom; just as Sanzio, Buonarotti, Sarpi, Reni, Vecelli, are universally known by their Christian names of Raphael, Michel Angelo, Fra Paolo, Guido, and Titian.

Several authors have followed Rossi in styling Galileo illegitimate, but without having any probable grounds even when they wrote, and the assertion has since been completely disproved by an inspection of the registers at Pisa and Florence, in which are preserved the dates of his birth, and of his mother's marriage, eighteen months previous to it.*

His father, Vincenzo Galilei, was a man of considerable talent and learning, with a competent knowledge of mathematies, and particularly devoted to the theory and practice of music, on which he published several esteemed treatises. The only one which it is at present easy to procure—his Dialogue on ancient and modern music—exhibits proofs, not only of a thorough acquaintance with his subject, but of a sound and vigorous understanding applied to other topics incidentally discussed. There is a passage in the introductory part, which becomes interesting when considered as affording some traces of the precepts by which Galileo was in all probability trained to reach his preeminent station in the intellectual world. "It appears to me," says one of the speakers in the dialogue, "that they who in proof of any assertion rely simply on the weight of authority, without adducing any argument in support of it, act very absurdly: I, on the contrary, wish to be allowed freely to question and freely to answer you without any sort of adulation, as well becomes those who are truly in search of truth." Sentiments like these were of rare occurrence at the close of the sixteenth century, and it is

---

* Power's Experimental Philosophy, 1663.

* Erythræus, Pinacotheca, vol. i.; Salusbury's Life of Galileo. Nelli, Vita di Gal. Galilei.

to be regretted that Vincenzo hardly lived long enough to witness his idea of a true philosopher splendidly realized in the person of his son. Vincenzo died at an advanced age, in 1591. His family consisted of three sons, Galileo, Michel Angelo, and Benedetto, and the same number of daughters, Giulia, Virginia, and Livia. After Vincenzo's death the chief support of the family devolved upon Galileo, who seems to have assisted them to his utmost power. In a letter to his mother, dated 1600, relative to the intended marriage of his sister Livia with a certain Pompeo Baldi, he agrees to the match, but recommends its temporary postponement, as he was at that time exerting himself to furnish money to his brother Michel Angelo, who had received the offer of an advantageous settlement in Poland. As the sum advanced to his brother, which prevented him from promoting his sister's marriage, did not exceed 200 crowns, it may be inferred that the family were in a somewhat straitened condition. However he promises, as soon as his brother should repay him, " to take measures for the young lady, since she too is bent upon *coming out* to prove the miseries of this world." —As Livia was at the date of this letter in a convent, the last expression seems to denote that she had been destined to take the veil. This proposed marriage never took place, but Livia was afterwards married to Taddeo Galletti: her sister Virginia married Benedetto Landucci. Galileo mentions one of his sisters, (without naming her) as living with him in 1619 at Bellosguardo. Michel Angelo is probably the same brother of Galileo who is mentioned by Liceti as having communicated from Germany some observations on natural history.* He finally settled in the service of the Elector of Bavaria; in what situation is not known, but upon his death the Elector granted a pension to his family, who then took up their abode at Munich. On the taking of that city in 1636, in the course of the bloody thirty years' war, which was then raging between the Austrians and Swedes, his widow and four of his children were killed, and every thing which they possessed was either burnt or carried away. Galileo sent for his two nephews, Alberto and a younger brother, to Arcetri near Florence, where he was then living. These two were then the only survivors of Michel Angelo's family; and many of Galileo's letters about that date contain allusions to the assistance he had been affording them. The last trace of Alberto is on his return into Germany to the Elector, in whose service his father had died. These details include almost every thing which is known of the rest of Vincenzo's family.

Galileo exhibited early symptoms of an active and intelligent mind, and distinguished himself in his childhood by his skill in the construction of ingenious toys and models of machinery, supplying the deficiencies of his information from the resources of his own invention; and he conciliated the universal good-will of his companions by the ready good nature with which he employed himself in their service and for their amusement. It is worthy of observation, that the boyhood of his great follower Newton, whose genius in many respects so closely resembled his own, was marked by a similar talent. Galileo's father was not opulent, as has been already stated: he was burdened with a large family, and was unable to provide expensive instructors for his son; but Galileo's own energetic industry rapidly supplied the want of better opportunities; and he acquired, under considerable disadvantages, the ordinary rudiments of a classical education, and a competent knowledge of the other branches of literature which were then usually studied. His leisure hours were applied to music and drawing; for the former accomplishment he inherited his father's talent, being an excellent performer on several instruments, especially on the lute; this continued to be a favourite recreation during the whole of his life. He was also passionately fond of painting, and at one time he wished to make it his profession: and his skill and judgment of pictures were highly esteemed by the most eminent contemporary artists, who did not scruple to own publicly their deference to young Galileo's criticism.

When he had reached his nineteenth year, his father, becoming daily more sensible of his superior genius, determined, although at a great personal sacrifice, to give him the advantages of an university education. Accordingly, in 1581,' he commenced his academical studies' in the university of his native town, Pisa, his father at this time intending that

---

* De his quæ diu vivunt. Patavii, 1612.

he should adopt the profession of medicine. In the matriculation lists at Pisa, he is styled Galileo, the son of Vincenzo Galilei, a Florentine, Scholar in Arts. It is dated 5th November, 1581. Viviani, his pupil, friend, and panegyrist, declares that, almost from the first day of his being enrolled on the lists of the academy, he was noticed for the reluctance with which he listened to the dogmas of the Aristotelian philosophy, then universally taught; and he soon became obnoxious to the professors from the boldness with which he promulgated what they styled his philosophical paradoxes. His early habits of free inquiry were irreconciteable with the mental quietude of his instructors, whose philosophic doubts, when they ventured to entertain any, were speedily lulled by a quotation from Aristotle. Galileo thought himself capable of giving the world an example of a sounder and more original mode of thinking; he felt himself destined to be the founder of a new school of rational and experimental philosophy. Of this we are now securely enjoying the benefits; and it is difficult at this time fully to appreciate the obstacles which then presented themselves to free inquiry: but we shall see, in the course of this narrative, how arduous their struggle was who happily effected this important revolution. The vindictive rancour with which the partisans of the old philosophy never ceased to assail Galileo is of itself a sufficient proof of the prominent station which he occupied in the contest.

Galileo's earliest mechanical discovery, to the superficial observer apparently an unimportant one, occurred during the period of his studies at Pisa. His attention was one day arrested by the vibrations of a lamp swinging from the roof of the cathedral, which, whether great or small, seemed to recur at equal intervals. The instruments then employed for measuring time were very imperfect: Galileo attempted to bring his observation to the test before quitting the church, by comparing the vibrations with the beatings of his own pulse, and his mind being then principally employed upon his intended profession, it occurred to him, when he had further satisfied himself of their regularity by repeated and varied experiments, that the process he at first adopted might be reversed, and that an instrument on this principle might be usefully employed in ascertaining the rate of the pulse, and its variation from day to day. He immediately carried the idea into execution, and it was for this sole and limited purpose that the first pendulum was constructed. Viviani tells us, that the value of the invention was rapidly appreciated by the physicians of the day, and was in common use in 1654, when he wrote.

Santorio, who was professor of medicine at Padua, has given representations of four different forms of these

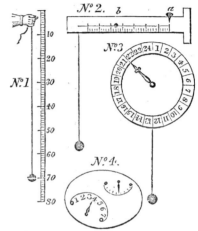

instruments, which he calls pulsilogies, (*pulsilogias*,) and strongly recommends to medical practitioners.* These instruments seem to have been used in the following manner: No. 1 consists merely of a weight fastened to a string and a graduated scale. The string being gathered up into the hand till the vibrations of the weight coincided with the beatings of the patient's pulse, the length was ascertained from the scale, which, of course, if great, indicated a languid, if shorter, a more lively action. In No. 2 the improvement is introduced of connecting the scale and string, the length of the latter is regulated by the turns of a peg at *a*, and a bead upon the string at *b* showed the measure. No. 3 is still more compact, the string being shortened by winding upon an axle at the back of the dial-plate. The construction of No. 4, which Santorio claims as his own improvement, is not given, but it is probable that the principal index, by its motion, shifted a weight to different distances from the point of suspension, and that the period of vibration

---

* Comment. in Avicennam. Venetiis, 1625.

was still more accurately adjusted by a smaller weight connected with the second index. Venturi seems to have mistaken the third figure for that of a pendulum clock, as he mentions this as one of the earliest adaptations of Galileo's principle to that purpose* ; but it is obvious, from Santorio's description, that it is nothing more than a circular scale, the index showing, by the figure to which it points, the length of string remaining unwound upon the axis. We shall, for the present, postpone the consideration of the invention of pendulum clocks, and the examination of the different claims to the honour of their first construction.

At the time of which we are speaking, Galileo was entirely ignorant of mathematics, the study of which was then at a low ebb, not only in Italy, but in every part of Europe. Commandine had recently revived a taste for the writings of Euclid and Archimedes, and Vieta Tartalea and others had made considerable progress in algebra, Guido Ubaldi and Benedetti had done something towards establishing the principles of statics, which was the only part of mechanics as yet cultivated; but with these inconsiderable exceptions the application of mathematics to the phenomena of nature was scarcely thought of. Galileo's first inducement to acquire a knowledge of geometry arose from his partiality for drawing and music, and from the wish to understand their principles and theory. His father, fearful lest he should relax his medical studies, refused openly to encourage him in this new pursuit ; but he connived at the instruction which his son now began to receive in the writings of Euclid, from the tuition of an intimate friend, named Ostilio Ricci, who was one of the professors in the university. Galileo's whole attention was soon directed to the enjoyment of the new sensations thus communicated to him, insomuch that Vincenzo, finding his prognostics verified, began to repent his indirect sanction, and privately requested Ricci to invent some excuse for discontinuing his lessons. But it was fortunately too late ; the impression was made and could not be effaced ; from that time Hippocrates and Galen lay unheeded before the young physician, and served only to conceal from his father's sight the mathematical volumes on which the whole of his time was really employed. His progress soon revealed the true nature of his pursuits: Vincenzo yielded to the irresistible predilection of his son's mind, and no longer attempted to turn him from the speculations to which his whole existence was thenceforward abandoned.

After mastering the elementary writers, Galileo proceeded to the study of Archimedes, and, whilst perusing the Hydrostatics of that author, composed his earliest work,—an Essay on the Hydrostatical Balance. In this he explains the method probably adopted by Archimedes for the solution of Hiero's celebrated question*, and shows himself already well acquainted with the true principles of specific gravities. This essay had an immediate and important influence on young Galileo's fortunes, for it introduced him to the approving notice of Guido Ubaldi, then one of the most distinguished mathematicians of Italy. At his suggestion Galileo applied himself to consider the position of the centre of gravity in solid bodies, a choice of subject that sufficiently showed the estimate Ubaldi had formed of his talents; for it was a question on which Commandine had recently written, and which engaged at that time the attention of geometricians of the highest order. Galileo tells us himself that he discontinued these researches on meeting with Lucas Valerio's treatise on the same subject. Ubaldi was so much struck with the genius displayed in the essay with which Galileo furnished him, that he introduced him to his brother, the Cardinal Del Monte: by this latter he was mentioned to Ferdinand de' Medici, the reigning Duke of Tuscany, as a young man of whom the highest expectations might be entertained. By the Duke's patronage he was nominated, in 1589, to the lectureship of mathematics at Pisa, being then in his twenty-sixth year. His public salary was fixed at the insignificant sum of sixty crowns annually, but he had an opportunity of greatly adding to his income by private tuition.

CHAPTER III.

Galileo at Pisa—Aristotle—Leonardo da Vinci—Galileo becomes a Copernican—Urstisius—Bruno—Experiments on falling bodies—Galileo at Padua—Thermometer.

No sooner was Galileo settled in his new office than he renewed his inquiries into the phenomena of nature with increased diligence. He instituted a course

---

* Essai sur les Ouvrages de Leonard da Vinci. Paris, 1797.

* See Treatise on HYDROSTATICS.

of experiments for the purpose of putting to the test the mechanical doctrines of Aristotle, most of which he found unsupported even by the pretence of experience. It is to be regretted that we do not more frequently find detailed his method of experimenting, than occasionally in the course of his dialogues, and it is chiefly upon the references which he makes to the results with which the experiments furnished him, and upon the avowed and notorious character of his philosophy, that the truth of these accounts must be made to depend. Venturi has found several unpublished papers by Galileo on the subject of motion, in the Grand Duke's private library at Florence, bearing the date of 1590, in which are many of the theorems which he afterwards developed in his Dialogues on Motion. These were not published till fifty years afterwards, and we shall reserve an account of their contents till we reach that period of his life.

Galileo was by no means the first who had ventured to call in question the authority of Aristotle in matters of science, although he was undoubtedly the first whose opinions and writings produced a very marked and general effect. Nizzoli, a celebrated scholar who lived in the early part of the 16th century, had condemned Aristotle's philosophy, especially his Physics, in very unequivocal and forcible terms, declaring that, although there were many excellent truths in his writings, the number was scarcely less of false, useless, and ridiculous propositions*. About the time of Galileo's birth, Benedetti had written expressly in confutation of several propositions contained in Aristotle's mechanics, and had expounded in a clear manner some of the doctrines of statical equilibrium.† Within the last forty years it has been established that the celebrated painter Leonardo da Vinci, who died in 1519, amused his leisure hours in scientific pursuits; and many ideas appear to have occurred to him which are to be found in the writings of Galileo at a later date. It is not impossible (though there are probably no means of directly ascertaining the fact) that Galileo may have been acquainted with Leonardo's investigations, although they remained, till very lately, almost unknown to the mathematical world. This supposition is rendered more probable from the fact, that Mazenta, the preserver of Leonardo's manuscripts, was, at the very time of their discovery, a contemporary student with Galileo at Pisa. Kopernik, or, as he is usually called, Copernicus, a native of Thorn in Prussia, had published his great work, De Revolutionibus, in 1543, restoring the knowledge of the true theory of the solar system, and his opinions were gradually and silently gaining ground.

It is not satisfactorily ascertained at what period Galileo embraced the new astronomical theory. Gerard Voss attributes his conversion to a public lecture of Mæstlin, the instructor of Kepler; and later writers (among whom is Laplace) repeat the same story, but without referring to any additional sources of information, and in most instances merely transcribing Voss's words, so as to shew indisputably whence they derived their account. Voss himself gives no authority, and his general inaccuracy makes his mere word not of much weight. The assertion appears, on many accounts, destitute of much probability. If the story were correct, it seems likely that some degree of acquaintance, if not of friendly intercourse, would have subsisted between Mæstlin, and his supposed pupil, such as in fact we find subsisting between Mæstlin and his acknowledged pupil Kepler, the devoted friend of Galileo; but, on the contrary, we find Mæstlin writing to Kepler himself of Galileo as an entire stranger, and in the most disparaging terms. If Mæstlin could lay claim to the honour of so celebrated a disciple, it is not likely that he could fail so entirely to comprehend the distinction it must confer upon himself as to attempt diminishing it by underrating his pupil's reputation. There is a passage in Galileo's works which more directly controverts the claim advanced for Mæstlin, although Salusbury, in his life of Galileo, having apparently an imperfect recollection of its tenor, refers to this very passage in confirmation of Voss's statement. In the second part of the dialogue on the Copernican system, Galileo makes Sagredo, one of the speakers in it, give the following account:—" Being very young, and having scarcely finished my course of philosophy, which I left off as being set upon other employments, there chanced to come into these parts a certain foreigner of Rostoch, *whose name, as I remember, was Christianus Ursti-sius*, a follower of Copernicus, who, in an academy, gave two or three lectures upon this point, to whom many flocked as auditors; but I, thinking they went

---

* Antibarbarus Philosophicus. Francofurti, 1674.
† Speculationum liber. Venetiis, 1585.

more for the novelty of the subject than otherwise, did not go to hear him; for I had concluded with myself that that opinion could be no other than a solemn madness; and questioning some of those who had been there, I perceived they all made a jest thereof, except one, who told me that the business was not altogether to be laughed at: and because the man was reputed by me to be very intelligent and wary, I repented that I was not there, and began from that time forward, as oft as I met with any one of the Copernican persuasion, to demand of them if they had been always of the same judgment. Of as many as I examined I found not so much as one who told me not that he had been a long time of the contrary opinion, but to have changed it for this, as convinced by the strength of the reasons proving the same; and afterwards questioning them one by one, to see whether they were well possessed of the reasons of the other side, I found them all to be very ready and perfect in them, so that I could not truly say that they took this opinion out of ignorance, vanity, or to show the acuteness of their wits. On the contrary, of as many of the Peripatetics and Ptolemeans as I have asked, (and out of curiosity I have talked with many,) what pains they had taken in the book of Copernicus, I found very few that had so much as superficially perused it, but of those who I thought had understood the same, not one: and, moreover, I have inquired amongst the followers of the Peripatetic doctrine, if ever any of them had held the contrary opinion, and likewise found none that had. Whereupon, considering that there was no man who followed the opinion of Copernicus that had not been first on the contrary side, and that was not very well acquainted with the reasons of Aristotle and Ptolemy, and, on the contrary, that there was not one of the followers of Ptolemy that had ever been of the judgment of Copernicus, and had left that to embrace this of Aristotle;—considering, I say, these things, I began to think that one who leaveth an opinion imbued with his milk and followed by very many, to take up another, owned by very few, and denied by all the schools, and that really seems a great paradox, must needs have been moved, not to say forced, by more powerful reasons. For this cause I am become very curious to dive, as they say, into the bottom of this business." It seems improbable that Galileo should think it worth while to give so detailed an account of the birth and growth of opinion in any one besides himself; and although Sagredo is not the personage who generally in the dialogue represents Galileo, yet as the real Sagredo was a young nobleman, a pupil of Galileo himself, the account cannot refer to him. The circumstance mentioned of the intermission of his philosophical studies, though in itself trivial, agrees very well with Galileo's original medical destination. Urstisius is not a fictitious name, as possibly Salusbury may have thought, when alluding to this passage; he was mathematical professor at Bâle, about 1567, and several treatises by him are still extant. In 1568 Voss informs us that he published some new questions on Purbach's Theory of the Planets. He died at Bâle in 1588, when Galileo was about twenty-two years old.

It is not unlikely that Galileo also, in part, owed his emancipation from popular prejudices to the writings of Giordano Bruno, an unfortunate man, whose unsparing boldness in exposing fallacies and absurdities was rewarded by a judicial murder, and by the character of heretic and infidel, with which his executioners endeavoured to stigmatize him for the purpose of covering over their own atrocious crime. Bruno was burnt at Rome in 1600, but not, as Montucla supposes, on account of his ". Spaccio della Bestia trionfante." The title of this book has led him to suppose that it was directed against the church of Rome, to which it does not in the slightest degree relate. Bruno attacked the fashionable philosophy alternately with reason and ridicule, and numerous passages in his writings, tedious and obscure as they generally are, show that he had completely outstripped the age in which he lived. Among his astronomical opinions, he believed that the universe consisted of innumerable systems of suns with assemblages of planets revolving round each of them, like our own earth, the smallness of which, alone, prevented their being observed by us. He remarked further, "that it is by no means improbable that there are yet other planets revolving round our own sun, which we have not yet noticed, either on account of their minute size or too remote distance from us." He declined asserting that all the apparently fixed stars are really so, considering this as not sufficiently proved, "because at such enormous distances the motions become difficult to estimate, and it is only by long observation that we can determine if any of these move round each other, or what other motions they may have." He ridiculed the Aristotelians in no very measured terms—" They harden themselves, and heat themselves, and embroil themselves for Aristotle; they call themselves his champions, they hate all but Aristotle's friends, they are ready to live and die for Aristotle, and yet they do not understand so much as the titles of Aristotle's chapters." And in another place he introduces an Aristotelian inquiring, " Do you take Plato for an ignoramus—Aristotle for an ass?" to whom he answers, " My son, I neither call them asses, nor you mules,—them baboons, nor you apes,—as you would have me: I told you that I esteem them the heroes of the world, but I will not credit them without sufficient reason; and if you were not both blind and deaf, you would understand that I must disbelieve their absurd and contradictory assertions."* Bruno's works, though in general considered those of a visionary and madman, were in very extensive circulation, probably not the less eagerly sought after from being included among the books prohibited by the Romish church;¹ and although it has been reserved for later observations to furnish complete verification of his most daring speculations, yet there was enough, abstractedly taken, in the wild freedom of his remarks, to attract a mind like Galileo's; and it is with more satisfaction that we refer the formation of his opinions to a man of undoubted though eccentric genius, like Bruno, than to such as Maestlin, who, though a diligent and careful observer, seems seldom to have taken any very enlarged views of the science on which he was engaged.

With a few exceptions similar to those above mentioned, the rest of Galileo's contemporaries well deserved the contemptuous epithet which he fixed on them of Paper Philosophers, for, to use his own words, in a letter to Kepler on this subject, " this sort of men fancied philosophy was to be studied like the Æneid or Odyssey, and that the true reading of nature was to be detected by the collation of texts." Galileo's own method of philosophizing was widely different; seldom omitting to bring with every new assertion the test of experiment, either directly in confirmation of it, or tending to show its probability and consistency. We have already seen that he engaged in a series of experiments to investigate the truth of some of Aristotle's positions. As fast as he succeeded in demonstrating the falsehood of any of them, he denounced them from his professorial chair with an energy and success which irritated more and more against him the other members of the academic body.

There seems something in the stubborn opposition which he encountered in establishing the truth of his mechanical theorems, still more stupidly absurd than in the ill will to which, at a later period of his life, his astronomical opinions exposed him: "it is intelligible that the vulgar should withhold their assent from one who pretended to discoveries in the remote heavens, which few possessed instruments to verify, or talents to appreciate; but it is difficult to find terms for stigmatizing the obdurate folly of those who preferred the evidence of their books to that of their senses, in judging of phenomena so obvious as those, for instance, presented by the fall of bodies to the ground. Aristotle had asserted, that if two different weights of the same material were let fall from the same height, the heavier one would reach the ground sooner than the other, in the proportion of their weights. The experiment is certainly not a very difficult one, but nobody thought of that method of argument, and consequently this assertion had been long received, upon his word, among the axioms of the science of motion. Galileo ventured to appeal from the authority of Aristotle to that of his own senses, and maintained that, with the exception of an inconsiderable difference, which he attributed to the disproportionate resistance of the air, they would fall in the same time. The Aristotelians ridiculed and refused to listen to such an idea. Galileo repeated his experiments in their presence from the famous leaning tower at Pisa: and with the sound of the simultaneously falling weights still ringing in their ears, they could persist in gravely maintaining that a weight of ten pounds would reach the ground in a tenth part of the time taken by one of a single pound, because they were able to quote chapter and verse in which Aristotle assures them that such is the fact. A temper of mind like this could not fail to produce ill will towards him who felt no scruples in exposing their wilful folly; and the watchful malice of these men soon found the means of making Galileo desirous of quitting

* De l'Infinito Universo. Dial. 3. La Cena de le Cenere, 1584.

his situation at Pisa. Don Giovanni de' Medici, a natural son of Cosmo, who possessed a slight knowledge of mechanics on which he prided himself, had proposed a contrivance for cleansing the port of Leghorn, on the efficiency of which Galileo was consulted. His opinion was unfavourable, and the violence of the inventor's disappointment, (for Galileo's judgment was verified by the result,) took the somewhat unreasonable direction of hatred towards the man whose penetration had foreseen the failure. Galileo's situation was rendered so unpleasant by the machinations of this person, that he decided on accepting overtures elsewhere, which had already been made to him; accordingly, under the negotiation of his staunch friend Guido Ubaldi, and with the consent of Ferdinand, be procured from the republic of Venice a nomination for six years to the professorship of mathematics in the university of Padua, whither he removed in September 1592.

Galileo's predecessor in the mathematical chair at Padua was Moleti, who died in 1588, and the situation had remained unfilled during the intervening four years. This seems to show that the directors attributed but little importance to the knowledge which it was the professor's duty to impart. This inference is strengthened by the fact, that the amount of the annual salary attached to it did not exceed 180 florins, whilst the professors of philosophy and civil law, in the same university, were rated at the annual stipends of 1400 and 1680 florins.* Galileo joined the university about a year after its triumph over the Jesuits, who had established a school in Padua about the year 1542, and, increasing yearly in influence, had shown symptoms of a design to get the whole management of the public education into the hands of their own body.† After several violent disputes it was at length decreed by the Venetian senate, in 1591, that no Jesuit should be allowed to give instruction at Padua in any of the sciences professed in the university. It does not appear that after this decree they were again troublesome to the university, but this first decree against them was followed, in 1606, by a second more peremptory, which banished them entirely from the Venetian territory. Galileo would of course find his fellow-professors much embittered against that society, and would naturally feel inclined to make common cause with them, so that it is not unlikely that the hatred which the Jesuits afterwards bore to Galileo on personal considerations, might be enforced by their recollection of the university to which he had belonged.

Galileo's writings now began to follow each other with great rapidity, but he was at this time apparently so careless of his reputation, that many of his works and inventions, after a long circulation in manuscript among his pupils and friends, found their way into the hands of those who were not ashamed to publish them as their own, and to denounce Galileo's claim to the authorship as the pretence of an impudent plagiarist. He was, however, so much beloved and esteemed by his friends, that they vied with each other in resenting affronts of this nature offered to him, and in more than one instance he was relieved, by their full and triumphant answers, from the trouble of vindicating his own character.

To this epoch of Galileo's life may be referred his re-invention of the thermometer. The original idea of this useful instrument belongs to the Greek mathematician Hero; and Santorio himself, who has been named as the inventor by Italian writers, and at one time claimed it himself, refers it to him. In 1638, Castelli wrote to Cesarmi that " he remembered an experiment shown to him more than thirty-five years back by Galileo, who took a small glass bottle, about the size of a hen's egg, the neck of which was twenty-two inches long, and as narrow as a straw. Having well heated the bulb in his hands, and then introducing its mouth into a vessel in which was a little water, and withdrawing the heat of his hand from the bulb, the water rose in the neck of the bottle more than eleven inches above the level in the vessel, and Galileo employed this principle in the construction of an instrument for measuring heat and cold."* In 1613, a Venetian nobleman named Sagredo, who has been already mentioned as Galileo's friend and pupil, writes to him in the following words: " I have brought the instrument which you invented for measuring heat into several convenient and perfect forms, so that the difference of temperature between two rooms is seen as far as 100 de-

---

* Riccoboni, Commentarii de Gymnasio PataVino, 1598.
† Nelli.

* Nelli.

grees."* This date is anterior to the claims both of Santorio and Drebbel, a Dutch physician, who was the first to introduce it into Holland.

Galileo's thermometer, as we have just seen, consisted merely of a glass tube ending in a bulb, the air in which, being partly expelled by heat, was replaced by water from a glass into which the open end of the tube was plunged, and the different degrees of temperature were indicated by the expansion of the air which yet remained in the bulb, so that the scale would be the reverse of that of the thermometer now in use, for the water would stand at the highest level in the coldest weather. It was, in truth, a barometer also, in consequence of the communication between the tube and external air, although Galileo did not intend it for this purpose, and when he attempted to determine the relative weight of the air, employed a contrivance still more imperfect than this rude barometer would have been. A passage among his posthumous fragments intimates that he subsequently used spirit of wine instead of water.

Viviani attributes an improvement of this imperfect instrument, but without specifying its nature, to Ferdinand II., a pupil and subsequent patron of Galileo, and, after the death of his father Cosmo, reigning duke of Florence. It was still further improved by Ferdinand's younger brother, Leopold de' Medici, who invented the modern process of expelling all the air from the tube by boiling the spirit of wine in it, and of hermetically sealing the end of the tube, whilst the contained liquid is in this expanded state, which deprived it of its barometrical character, and first made it an accurate thermometer. The final improvement was the employment of mercury instead of spirit of wine, which is recommended by Lana so early as 1670, on account of its equable expansion.† For further details on the history and use of this instrument, the reader may consult the Treatises on the THERMOMETER and PYROMETER.

CHAPTER IV.

Astronomy before Copernicus—Fracastoro — Bacon — Kepler — Galileo's Treatise on the Sphere.

THIS period of Galileo's lectureship at Padua derives interest from its including the first notice which we find of his having embraced the doctrines of the Copernican astronomy. Most of our readers are aware of the principles of the theory of the celestial motions which Copernicus restored; but the number of those who possess much knowledge of the cumbrous and unwieldy system which it superseded is perhaps more limited. The present is not a fit opportunity to enter into many details respecting it; these will find their proper place in the History of Astronomy: but a brief sketch of its leading principles is necessary to render what follows intelligible.

The earth was supposed to be immoveably fixed in the centre of the universe, and immediately surrounding it the atmospheres of air and fire, beyond which the sun, moon, and planets, were thought to be carried round the earth, fixed each to a separate orb or heaven of solid but transparent matter. The order of distance in which they were supposed to be placed with regard to the central earth was as follows: The Moon, Mercury, Venus, The Sun, Mars, Jupiter, and Saturn. It became a question in the ages immediately preceding Copernicus, whether the Sun was not nearer the Earth than Mercury, or at least than Venus; and this question was one on which the astronomical theorists were then chiefly divided.

We possess at this time a curious record of a former belief in this arrangement of the Sun and planets, in the order in which the days of the week have been named from them. According to the dreams of Astrology, each planet was supposed to exert its influence in succession, reckoning from the most distant down to the nearest, over each hour of the twenty-four. The planet which was supposed to predominate over the first hour, gave its name to that day.* The general reader will trace this curious fact more easily with the French or Latin names than with the English, which have been translated into the titles of the corresponding Saxon deities. Placing the Sun and planets in the following order, and beginning, for instance, with Monday, or the Moon's day; Saturn ruled the second hour of that day, Jupiter the third, and so round till we come again and again to the Moon on the 8th, 15th, and 22d hours; Saturn ruled the 23d,

---

* Venturi. Memorie e Lettere di Gal. Galilei. Modena, 1821.
† Prodromo all' Arte Maestra. Brescia, 1670.

* Dion Cassius, lib. 37.

Jupiter the 24th, so that the next day would be the day of Mars, or, as the Saxons translated it, Tuisco's day, or Tuesday. In the same manner the following days would belong respectively to Mercury or Woden, Jupiter or Thor, Venus or Frea, Saturn or Seater, the Sun, and again the Moon. In this manner the whole week will be found to complete the cycle of the seven planets.

The other stars were supposed to be fixed in an outer orb, beyond which were two crystalline spheres, (as they were called,) and on the outside of all, the *primum mobile* or *first moveable*, which sphere was supposed to revolve round the earth in twenty-four hours, and by its friction, or rather, as most of the philosophers of that day chose to term it, by the sort of heavenly influence which it exercised on the interior orbs, to carry them round with a similar motion. Hence the diversity of day and night. But beside this principal and general motion, each orb was supposed to have one of its own, which was intended to account for the apparent changes of position of the planets with respect to the fixed stars and to each other. This supposition, however, proving insufficient to account for all the irregularities of motion observed, two hypotheses were introduced.—First, that to each planet belonged several concentric spheres or heavens, casing each other like the coats of an onion, and, secondly, that the centres of these solid spheres, with which the planet revolved, were placed in the circumference of a secondary revolving sphere, the centre of which secondary sphere was situated at the earth. They thus acquired the names of Eccentrics or Epicycles, the latter word signifying a circle upon a circle. The whole art of astronomers was then directed towards inventing and combining different eccentric and epicyclical motions, so as to represent with tolerable fidelity the ever varying phenomena of the heavens. Aristotle had lent his powerful assistance in this, as in other branches of natural philosophy, in enabling the false system to prevail against and obliterate the knowledge of the true, which, as we gather from his own writings, was maintained by some philosophers before his time. Of these ancient opinions, only a few traces now remain, principally preserved in the works of those who were adverse to them. Archimedes says expressly that Aristarchus of Samos, who lived about 300 B. C., taught the immobility of the sun and stars, and that the earth is carried round the central sun.* Aristotle's words are: " Most of those who assert that the whole concave is finite, say that the earth is situated in the middle point of the universe: those who are called Pythagoreans, who live in Italy, are of a contrary opinion. For they say that fire is in the centre, and that the earth, which, according to them, is one of the stars, occasions the change of day and night by its own motion, with which it is carried about the centre." It might be doubtful, upon this passage alone, whether the Pythagorean theory embraced more than the diurnal motion of the earth, but a little farther, we find the following passage: " Some, as we have said, make the earth to be one of the stars: others say that it is placed in the centre of the Universe, and revolves on a central axis."† From

---

* The pretended translation by Roberval of an Arabic version of Aristarchus, " De Systemate Mundi," in which the Copernican system is fully developed, is spurious. Menage asserts this in his observations on Diogen Laert. lib. 8, sec. 85, tom. ii., p. 389. (Ed. Amst. 1692.) The commentary contains many authorities well worth consulting. Delambre, Histoire de l'Astronomie, infers it from its not containing some opinions which Archimedes tells us were held by Aristarchus. A more direct proof may be gathered from the following blunder of the supposed translator. Astronomers had been long aware that the earth in different parts of her orbit is at different distances from the sun. Roberval wished to claim for Aristarchus the credit of having known this, and introduced into his book, not only the mention of the fact, but an explanation of its cause. Accordingly he makes Aristarchus give a reason " why the sun's apogee (or place of greatest distance from the earth) must always be at the north summer solstice." In fact, it was there, or nearly so, in Roberval's time, and he knew not but that it had always been there. It is however moveable, and, when Aristarchus lived, was nearly half way between the solstices and equinoxes. He therefore would hardly have given a reason for the necessity of a phenomenon of which, if he observed anything on the subject, he must have observed the contrary. The change in the obliquity of the earth's axis to the ecliptic was known in the time of Roberval, and he accordingly has introduced the proper value which it had in Aristarchus's time.

† De Cœlo. lib. 2.

which, in conjunction with the former extract, it very plainly appears that the Pythagoreans maintained both the diurnal and annual motions of the earth.

Some idea of the supererogatory labour entailed upon astronomers by the adoption of the system which places the earth in the centre, may be formed in a popular manner by observing, in passing through a thickly planted wood, in how complicated a manner the relative positions of the trees appear at each step to be continually changing, and by considering the difficulty with which the laws of their apparent motions could be traced, if we were to attempt to refer these changes to a real motion of the trees instead of the traveller. The apparent complexity in the heavens is still greater than in the case suggested; because, in addition to the earth's motions, with which all the stars appear to be impressed, each of the planets has also a real motion of its own, which of course greatly contributes to perplex and complicate the general appearances. Accordingly the heavens rapidly became, under this system,

" With centric and eccentric scribbled o'er,
Cycle and epicycle, orb in orb;"*

crossing and penetrating each other in every direction. Maestlin has given a concise enumeration of the principal orbs which belonged to this theory. After warning the readers that "they are not mere fictions which have nothing to correspond with them out of the imagination, but that they exist really, and bodily in the heavens," he describes seven principal spheres belonging to each planet, which he classes as Eccentrics, Epicycles, and Concentrepicycles, and explains their use in accounting for the planet's revolutions, motions of the apogee, and nodes, &c. &c. In what manner this multitude of solid and crystalline orbs were secured from injuring or interfering with each other was not very closely inquired into.

The reader will cease to expect any very intelligible explanation of this and numberless other difficulties which belong to this unwieldy machinery, when he is introduced to the reasoning by which it was upheld. Gerolamo Fracastoro, who lived in the sixteenth century, writes in the following terms, in his work entitled Homocentrica, (certainly one of the best productions of the day,) in which he endeavours to simplify the necessary apparatus, and to explain all the phenomena (as the title of his book implies) by concentric spheres round the earth. " There are some, not only of the ancients but also among the moderns, who believe that the stars move freely without any such agency; but it is difficult to conceive in what manner they have imbued themselves with this notion, *since not only reason, but the very senses, inform us that all the stars are carried round fastened to solid spheres.*" What ideas Fracastoro entertained of the evidence of the "senses" it is not now easy to guess, but he goes on to give a specimen of the " reasoning" which appeared to him so incontrovertible. " The planets are observed to move one while forwards, then backwards, now to the right, now to the left, quicker and slower by turns; which variety is consistent with a compound structure like that of an animal, which possesses in itself various springs and principles of action, but is totally at variance with our notion of a simple and undecaying substance like the heavens and heavenly bodies. For that which is simple, is altogether single, and singleness is of one only nature, and one nature can be the cause of only one effect; and therefore it is altogether impossible that the stars of themselves should move with such variety of motion. And besides, if the stars move by themselves, they either move in an empty space, or in a fluid medium like the air. But there cannot be such a thing as empty space, and if there were such a medium, the motion of the star would occasion condensation and rarefaction in different parts of it, which is the property of corruptible bodies, and where they exist some violent motion is going on; but the heavens are incorruptible and are not susceptible of violent motion, and hence, and from many other similar reasons, any one who is not obstinate may satisfy himself that the stars cannot have any independent motion."

Some persons may perhaps think that arguments of this force are unnecessarily dragged from the obscurity to which they are now for the most part happily consigned; but it is essential, in order to set Galileo's character and merits in their true light, to show how low at this

* Paradise Lost, b. viii. V. 83.
† Itaque tam circulos primi motus quam orbes secundorum mobilium revera in cœlesti corpore esse concludimus, &c. Non ergo sunt mera figmenta, quibus extra mentem nihil correspondeat. M. Maestlini, De Astronomiæ Hypothesibus disputatio. Heidelbergæ, 1592.

time philosophy had fallen. For we shall form a very inadequate notion of his powers and deserts if we do not contemplate him in the midst of men who, though of undoubted talent and ingenuity, could so far bewilder themselves as to mistake such a string of unmeaning phrases for argument: we must reflect on the difficulty every one experiences in delivering himself from the erroneous impressions of infancy, which will remain stamped upon the imagination in spite of all the efforts of matured reason to erase them, and consider every step of Galileo's course as a triumph over difficulties of a like nature. We ought to be fully penetrated with this feeling before we sit down to the perusal of his works, every line of which will then increase our admiration of the penetrating acuteness of his invention and unswerving accuracy of his judgment. In almost every page we discover an allusion to some new experiment, or the germ of some new theory; and amid all this wonderful fertility it is rarely indeed that we find the exuberance of his imagination seducing him from the rigid path of philosophical induction. This is the more remarkable as he was surrounded by friends and contemporaries of a different temperament and much less cautious disposition. A disadvantageous contrast is occasionally furnished even by the sagacious Bacon, who could so far deviate from the sound principles of inductive philosophy, as to write, for instance, in the following strain, bordering upon the worst manner of the Aristotelians:—
" Motion in a circle has no limit, and seems to emanate from the appetite of the body, which moves only for the sake of moving, and that it may follow itself and seek its own embraces, and put in action and enjoy its own nature, and exercise its peculiar operation: on the contrary, motion in a straight line seems transitory, and to move towards a limit of cessation or rest, and that it may reach some point, and then put off its motion."* Bacon rejected all the machinery of the *primum mobile* and the solid spheres, the eccentrics and the epicycles, and carried his dislike of these doctrines so far as to assert that nothing short of their gross absurdity could have driven theorists to the extravagant supposition of the motion of the earth, which, said he, " we know to be most false."* Instances of extravagant suppositions and premature generalizations are to be found in almost every page of his other great contemporary, Kepler.

It is with pain that we observe Delambre taking every opportunity, in his admirable History of Astronomy, to undervalue and sneer at Galileo, seemingly for the sake of elevating the character of Kepler, who appears his principal favourite, but whose merit as a philosopher cannot safely be brought into competition with that of his illustrious contemporary. Delambre is especially dissatisfied with Galileo, for taking no notice, in his " System of the World," of the celebrated laws of the planetary motions which Kepler discovered, and which are now inseparably connected with his name. The analysis of Newton and his successors has now identified those apparently mysterious laws with the general phenomena of motion, and has thus entitled them to an attention of which, before that time, they were scarcely worthy; at any rate not more than is at present the empirical law which includes the distances of all the planets from the sun (roughly taken) in one algebraical formula. The observations of Kepler's day were scarcely accurate enough to prove that the relations which he discovered between the distances of the planets from the sun and the periods of their revolutions around him were necessarily to be received as demonstrated truths; and Galileo surely acted most prudently and philosophically in holding himself altogether aloof from Kepler's fanciful devices and numeral concinnities, although, with all the extravagance, they possessed much of the genius of the Platonic reveries, and although it did happen that Galileo, by systematically avoiding them, failed to recognise some important truths. Galileo probably was thinking of those very laws, when he said of Kepler, " He possesses a bold and free genius, perhaps too much so; but his mode of philosophizing is widely different from mine." We shall have further occasion in the sequel to recognise the justice of this remark.

In the treatise on the Sphere which bears Galileo's name, and which, if he be indeed the author of it, was composed during the early part of his residence at

---

* Opuscula Philosophica, Thema Cœli.

* " Nobis constat falsissimum esse." De Aug. Scient. lib. iii. c. 3. 1623.

Padua, he also adopts the Ptolemaic system, placing the earth immoveable in the centre, and adducing against its motion the usual arguments, which in his subsequent writings he ridicules and refutes. Some doubts have been expressed of its authenticity; but, however this may be, we have it under Galileo's own hand that he taught the Ptolemaic system, in compliance with popular prejudices, for some time after he had privately become a convert to the contrary opinions. In a letter, apparently the first which he wrote to Kepler, dated from Padua, 1597, he says, acknowledging the receipt of Kepler's Mysterium Cosmographicum, " I have as yet read nothing beyond the preface of your book, from which however I catch a glimpse of your meaning, and feel great joy on meeting with so powerful an associate in the pursuit of truth, and consequently such a friend to truth itself, for it is deplorable that there should be so few who care about truth, and who do not persist in their perverse mode of philosophizing; but as this is not the fit time for lamenting the melancholy condition of our times, but for congratulating you on your elegant discoveries in confirmation of the truth, I shall only add a promise to peruse your book dispassionately, and with a conviction that. I shall find in it much to admire. *This I shall do the more willingly because many years ago I became a convert to the opinions of Copernicus,** and by that theory have succeeded in fully explaining many phenomena, which on the contrary hypothesis are altogether inexplicable. I have arranged many arguments and confutations of the opposite opinions, which however I have not yet dared to publish*, fearing the fate of our master Copernicus, who, although he has earned immortal fame among a few, yet by an infinite number (for so only can the number of fools be measured) is exploded and derided. If there were many such as you, I would venture to publish my speculations; but, since that is not so, I shall take time to consider of it." This interesting letter was the beginning of the friendship of these two great men, which lasted uninterruptedly till 1632, the date of Kepler's death. That extraordinary genius never omitted an opportunity of testifying his admiration of Galileo, although there were not wanting persons envious of their good understanding, who exerted themselves to provoke coolness and quarrel between them. Thus Brutius writes to Kepler in 1602*: " Galileo tells me he has written to you, and has got your book, which however he denied to Magini, and I abused him for praising you with too many qualifications. I know it to be a fact that, both in his lectures, and elsewhere, he is publishing your inventions as his own; but I have taken care, and shall continue to do so, that all this shall redound not to his credit but to yours." The only notice which Kepler took of these repeated insinuations, which appear to have been utterly groundless, was, by renewed expressions of respect and admiration, to testify the value he set upon his friend and fellow-labourer in philosophy.

CHAPTER V.

Galileo re-elected Professor at Padua —New star—Compass of proportion—Capra—Gilbert—Proposals to return to Pisa—Lost writings—Cavalieri.

GALILEO's reputation was now rapidly increasing: his lectures were attended by many persons of the highest rank; among whom were the Archduke Ferdinand, afterwards Emperor of Germany, the Landgrave of Hesse, and the Princes of Alsace and Mantua. On the expiration of the first period for which he had been elected professor, he was rechosen for a similar period, with a salary increased to 320 florins. The immediate occasion of this augmentation is said by Fabroni†, to have arisen out of the malice of an ill wisher of Galileo, who, hoping to do him disservice, apprized the senate that he was not married to Marina Gamba, then living with him, and the mother of his son Vincenzo. Whether or not the senate might consider themselves entitled to inquire into the morality of his private life, it was probably from a wish to mark their sense of the informer's impertinence, that they returned the brief answer, that "if he had a family to provide for, he stood the more in need of an increased stipend."

During Galileo's residence at Padua, and, according to Viviani's intimation, towards the thirtieth year of his age, that is to say in 1594, he experienced

---

* Id autum eò libentius faciam, quod in Copernici sententiam multis abhinc annis venerim.—Kepl. Epistolæ.

* Kepleri Epistolæ.
† Vitæ Italorum Illustrium.

the first attack of a disease which pressed heavily on him for the rest of his life. He enjoyed, when a young man, a healthy and vigorous constitution, but chancing to sleep one afternoon near an open window, through which was blowing a current of air cooled artificially by the fall of water, the consequences were most disastrous to him. He contracted a sort of chronic complaint, which showed itself in acute pains in his limbs, chest, and back, accompanied with frequent hæmorrhages and loss of sleep and appetite; and this painful disorder thenceforward never left him entirely, but recurred intermittingly, with greater or less violence, as long as he lived. Others of the party did not even escape so well, but died shortly after committing this imprudence.

In 1604, the attention of astronomers was called to the contemplation of a new star, which appeared suddenly with great splendour in the constellation Serpentarius, or Ophiuchus, as it is now more commonly called. Maestlin, who was one of the earliest to notice it, relates his observations in the following words: " How wonderful is this new star! I am certain that I did not see it before the 29th of September, nor indeed, on account of several cloudy nights, had I a good view till the 6th of October. Now that it is on the other side of the sun, instead of surpassing Jupiter as it did, and almost rivalling Venus, it scarcely matches the Cor Leonis, and hardly surpasses Saturn. It continues however to shine with the same bright and strongly sparkling light, and changes its colours almost with every moment; first tawny, then yellow, presently purple and red, and, when it has risen above the vapours, most frequently white." This was by no means an unprecedented phenomenon; and the curious reader may find in Riccioli* a catalogue of the principal new stars which have at different times appeared. There is a tradition of a similar occurrence as early as the times of the Greek astronomer Hipparchus, who is said to have been stimulated by it to the formation of his catalogue of the stars; and only thirty-two years before, in 1572, the same remarkable phenomenon in the constellation Cassiopeia was mainly instrumental in detaching the celebrated Tycho Brahe from the chemical studies, which till then divided his attention with astronomy. Tycho's star disappeared at the end of two years; and at that time Galileo was a child. On the present occasion, he set himself earnestly to consider the new phenomenon, and embodied the results of his observations in three lectures, which have been unfortunately lost. Only the exordium of the first has been preserved: in this he reproaches his auditors with their general insensibility to the magnificent wonders of creation daily exposed to their view, in no respect less admirable than the new prodigy, to hear an explanation of which they had hurried in crowds to his lecture room. He showed, from the absence of parallax, that the new star could not be, as the vulgar hypothesis represented, a mere meteor engendered in our atmosphere and nearer the earth than the moon, but must be situated among the most remote heavenly bodies. This was inconceivable to the Aristotelians, whose notions of a perfect, simple, and unchangeable sky were quite at variance with the introduction of any such new body; and we may perhaps, consider these lectures as the first public declaration of Galileo's hostility to the old Ptolemaic and Aristotelian astronomy.

In 1606 he was reappointed to the lectureship, and his salary a second time increased, being raised to 520 florins. His public lectures were at this period so much thronged that the ordinary place of meeting was found insufficient to contain his auditors, and he was on several occasions obliged to adjourn to the open air,—even from the school of medicine, which was calculated to contain one thousand persons.

About this time he was considerably annoyed by a young Milanese, of the name of Balthasar Capra, who pirated an instrument which Galileo had invented some years before, and had called the geometrical and military compass. The original offender was a German named Simon Mayer, whom we shall meet with afterwards arrogating to himself the merit of one of Galileo's astronomical discoveries; but on this occasion, as soon as he found Galileo disposed to resent the injury done to him, he hastily quitted Italy, leaving his friend Capra to bear alone the shame of the exposure which followed. The instrument is of simple construction, consisting merely of two straight rulers, connected by a joint; so that they can be set to any required angle. This simple and useful instrument, now called the Sector, is to be found in almost every

---

* Almagestum Novum, vol. i.

case of mathematical instruments. Instead of the trigonometrical and logarithmic lines which are now generally engraved upon it, Galileo's compass merely contained, on one side, three pairs of lines, divided in simple, duplicate, and triplicate proportion, with a fourth pair on which were registered the specific gravities of several of the most common metals. These were used for multiplications, divisions, and the extraction of roots; for finding the dimensions of equally heavy balls of different materials, &c. On the other side were lines contrived for assisting to describe any required polygon on a given line; for finding polygons of one kind equal in area to those of another; and a multitude of other similar operations useful to the practical engineer.

Unless the instrument, which is now called Gunter's scale, be much altered from what it originally was, it is difficult to understand on what grounds Salusbury charges Gunter with plagiarism from Galileo's Compass. He declares that he has closely compared the two, and can find no difference between them.* There has also been some confusion, by several writers, between this instrument and what is now commonly called the Proportional Compass. The latter consists of two slips of metal pointed at each end, and connected by a pin which, sliding in a groove through both, can be shifted to different positions. Its use is to find proportional lines; for it is obvious that the openings measured by each pair of legs will be in the same proportion in which the slips are divided by the centre. The divisions usually marked on it are calculated for finding the submultiples of straight lines, and the chords of submultiple arcs. Montucla has mentioned this mistake of one instrument for the other, and charges Voltaire with the more inexcusable error of confounding Galileo's with the Mariner's Compass. He refers to a treatise by Hulsius for his authority in attributing the Proportional Compass to Burg, a German astronomer of some celebrity. Horcher also has been styled the inventor; but he did no more than describe its form and application. In the frontispiece of his book is an engraving of this compass exactly similar to those which are now used.† To the description which Galileo published of his compass, he added a short treatise on the method of measuring heights and distances with the quadrant and plumb line. The treatise, which is printed by itself at the end of the first volume of the Padua edition of Galileo's works, contains nothing more than the demonstrations belonging to the same operations. They are quite elementary, and contain little or nothing that was new even at that time.

Such an instrument as Galileo's Compass was of much more importance before the grand discovery of logarithms than it can now be considered; however it acquires an additional interest from the value which he himself set on it. In 1607, Capra, at the instigation of Mayer, published as his own invention what he calls the proportional hoop, which is a mere copy of Galileo's instrument. This produced from Galileo a long essay, entitled "A Defence of Galileo against the Calumnies and Impostures of Balthasar Capra." His principal complaint seems to have been of the misrepresentations which Capra had published of his lectures on the new star already mentioned, but he takes occasion, after pointing out the blunders and falsehoods which Capra had committed on that occasion, to add a complete proof of his piracy of the geometrical compass. He showed, from the authenticated depositions of workmen, and of those for whom the instruments had been fabricated, that he had devised them as early as the year 1597, and had explained their construction and use both to Balthasar himself and to his father Aurelio Capra, who was then residing in Padua. He gives, in the same essay, the minutes of a public meeting between himself and Capra, in which he proved, to the satisfaction of the university, that wherever Capra had endeavoured to introduce into this book propositions which were not to be met with in Galileo's, he had fallen into the greatest absurdities, and betrayed the most complete ignorance of his subject. The consequence of this public exposure, and of the report of the famous Fra Paolo Sarpi, to whom the matter had been referred, was a formal prohibition by the university of Capra's publication, and all copies of the book then on hand were seized, and probably destroyed, though Galileo has preserved it from oblivion by incorporating it in his own publication.

Nearly at the same time, 1607, or immediately after, he first turned his attention towards the loadstone, on which our

* Math Coll. vol. ii.
† Constructio Circini Proportionum. Moguntiæ, 1605.

c

countryman Gilbert had already published his researches, conducted in the true spirit of the inductive method. Very little that is original is to be found in Galileo's works on this subject, except some allusions to his method of arming magnets, in which, as in most of his practical and mechanical operations, he appears to have been singularly successful. Sir Kenelm Digby* asserts, that the magnets armed by Galileo would support twice as great a weight as one of Gilbert's of the same size. Galileo was well acquainted, as appears from his frequent allusions in different parts of his works, with what Gilbert had done, of whom he says, " I extremely praise, admire, and envy this author;— I think him, moreover, worthy of the greatest praise for the many new and true observations that he has made to the disgrace of so many vain and fabling authors, who write, not from their own knowledge only, but repeat every thing they hear from the foolish vulgar, without attempting to satisfy themselves of the same by experience, perhaps that they may not diminish the size of their books."

Galileo's reputation being now greatly increased, proposals were made to him, in 1609, to return to his original situation at Pisa. He had been in the habit of passing over to Florence during the academic vacation, for the purpose of giving mathematical instruction to the younger members of Ferdinand's family; and Cosmo, who had now succeeded his father as duke of Tuscany, regretted that so masterly a genius had been allowed to leave the university which he naturally should have graced. A few extracts from Galileo's answers to these overtures will serve to show the nature of his situation at Padua, and the manner in which his time was there occupied. " I will not hesitate to say, having now laboured during twenty years, and those the best of my life, in dealing out, as one may say, in detail, at the request of any body, the little talent which God has granted to my assiduity in my profession, that my wish certainly would be to have sufficient rest and leisure to enable me, before my life comes to its close, to conclude three great works which I have in hand, and to publish them; which might perhaps bring some credit to me, and to those who had favoured me in this undertaking, and possibly may be of greater and more frequent service to students than in the rest of my life I could personally afford them. Greater leisure than I have here I doubt if I could meet with elsewhere, so long as I am compelled to support my family from my public and private lectures, (nor would I willingly lecture in any other city than this, for several reasons which would be long to mention) nevertheless not even the liberty I have here is sufficient, where I am obliged to spend many, and often the best hours of the day at the request of this and that man. —My public salary here is 520 florins, which I am almost certain will be advanced to as many crowns upon my re-election, and these I can greatly increase by receiving pupils, and from private lectures, to any extent that I please. My public duty does not confine me during more than 60 half hours in the year, and even that not so strictly but that I may, on occasion of any business, contrive to get some vacant days; the rest of my time is absolutely at my own disposal; but because my private lectures and domestic pupils are a great hindrance and interruption of my studies, I wish to live entirely exempt from the former, and in great measure from the latter: for if I am to return to my native country, I should wish the first object of his Serene Highness to be, that leisure and opportunity should be given me to complete my works without employing myself in lecturing. — And, in short, I should wish to gain my bread from my writings, which I would always dedicate to my Serene Master.—The works which I have to finish are principally —two books on the system or structure of the Universe, an immense work, full of philosophy, astronomy, and geometry; three books on Local Motion, a science entirely new, no one, either ancient or modern, having discovered any of the very many admirable accidents which I demonstrate in natural and violent motions, so that I may with very great reason call it a new science, and invented by me from its very first principles; three books of Mechanics, two on the demonstration of principles and one of problems; and although others have treated this same matter, yet all that has been hitherto written, neither in quantity, nor otherwise, is the quarter of what I am writing on it. I have also different treatises on natural subjects; On sound and speech; On light and colours; On the tide; On the composition of continuous quantity; On the

---

* Treatise of the Nature of Bodies. London, 1665.

motions of animals;—And others besides. I have also an idea of writing some books relating to the military art, giving not only a model of a soldier, but teaching with very exact rules every thing which it is his duty to know that depends upon mathematics; as the knowledge of castrametation, drawing up battalions, fortifications, assaults, planning, surveying, the knowledge of artillery, the use of instruments, &c. I also wish to reprint the 'Use of my Geometrical Compass,' which is dedicated to his highness, and which is no longer to be met with; for this instrument has experienced such favour from the public, that in fact no other instruments of this kind are now made, and I know that up to this time several thousands of mine have been made.—I say nothing as to the amount of my salary, feeling convinced that as I am to live upon it, the graciousness of his highness would not deprive me of any of those comforts, which, however, I feel the want of less than many others; and therefore I say nothing more on the subject. Finally, on the title and profession of my service, I should wish that to the name of Mathematician, his highness would add that of Philosopher, as I profess to have studied a greater number of years in philosophy than months in pure mathematics; and how I have profited by it, and if I can or ought to deserve this title, I may let their highnesses see as often as it shall please them to give me an opportunity of discussing such subjects in their presence with those who are most esteemed in this knowledge." It may perhaps be seen in the expressions of this letter, that Galileo was not inclined to undervalue his own merits, but the peculiar nature of the correspondence should be taken into account, which might justify his indulging a little more than usual in self-praise, and it would have been perhaps almost impossible for him to have remained entirely blind to his vast superiority over his contemporaries.

Many of the treatises which Galileo here mentions, as well as another on dialling, have been irrecoverably lost, through the superstitious weakness of some of his relations, who after his death suffered the family confessor to examine his papers, and to destroy whatever seemed to him objectionable; a portion which, according to the notions then prevalent, was like to comprise the most valuable part of the papers submitted to this expurgation. It is also supposed that many were burnt by his infatuated grandson Cosimo, who conceived he was thus offering a proper and pious sacrifice before devoting himself to the life of a missionary. A Treatise on Fortification, by Galileo, was found in 1793, and is contained among the documents published by Venturi. Galileo does not profess in it to give much original matter, but to lay before his readers a compendium of the most approved principles then already known. It has been supposed that Gustavus Adolphus of Sweden attended Galileo's lectures on this subject, whilst in Italy; but the fact is not satisfactorily ascertained. Galileo himself mentions a Prince Gustavus of Sweden to whom he gave instruction in mathematics, but the dates cannot well be made to agree. The question deserves notice only from its having been made the subject of controversy.

The loss of Galileo's Essay on Continuous Quantity is particularly to be regretted, as it would be highly interesting to see how far he succeeded in methodizing his thoughts on this important topic. It is to his pupil Cavalieri (who refused to publish his book so long as he hoped to see Galileo's printed) that we owe " The Method of Indivisibles," which is universally recognized as one of the first germs of the powerful methods of modern analysis. Throughout Galileo's works we find many indications of his having thought much on the subject, but his remarks are vague, and bear little, if at all, on the application of the method. To this the chief part of Cavalieri's book is devoted, though he was not so entirely regardless of the principles on which his method of measuring spaces is founded, as he is sometimes represented. This method consisted in considering lines as made up of an infinite number of points, surfaces in like manner as composed of lines, and solids of surfaces; but there is an observation at the beginning of the 7th book, which shews clearly that Cavalieri had taken a much more profound view of the subject than is implied in this superficial exposition, and had approached very closely to the apparently more exact theories of his successors. Anticipating the objections to his hypothesis, he argues, that " there is no necessity to suppose the continuous quantities made up of these indivisible parts, *but only that they will observe the same ratios as those parts do.*" It ought not to be omitted, that Kepler also had given an impulse to

Cavalieri in his "New method of Guaging," which is the earliest work with which we are acquainted, where principles of this sort are employed.*

Chapter VI.

Invention of the telescope—Fracastoro—Porta—Reflecting telescope—Roger Bacon—Digges—De Dominis—Jansen—Lipperhey—Galileo constructs telescopes—Microscopes—Re-elected Professor at Padua for life.

THE year 1609 was signalized by Galileo's discovery of the telescope, which, in the minds of many, is the principal, if not the sole invention associated with his name. It cannot be denied that his fame, as the founder of the school of experimental philosophy, has been in an unmerited degree cast into the shade by the splendour of his astronomical discoveries; yet Lagrange† surely errs in the opposite extreme, when he almost denies that these form any real or solid part of the glory of this great man; and Montucla‡ omits an important ingredient in his merit, when he (in other respects very justly) remarks, that it required far less genius to point a telescope towards the heavens than to trace the unheeded, because daily recurring, phenomena of motion up to its simple and primary laws. We are to remember that in the days of Galileo a telescope could scarcely be pointed to the heavens with impunity, and that a courageous mind was required to contradict, and a strong one to bear down, a party, who, when invited to look on any object in the heavens which Aristotle had never suspected, immediately refused all credit to those senses, to which, on other occasions, they so confidently appealed. It surely is a real and solid part of Galileo's glory that he consumed his life in laborious and indefatigable observations, and that he persevered in announcing his discoveries undisgusted by the invectives, and undismayed by the persecutions, to which they subjected him. Plagiarist! liar! impostor! heretic! were among the expressions of malignant hatred lavished upon him, and although he also was not without some violent and foul-mouthed partisans, yet it must be told to his credit that he himself seldom condescended to notice these torrents of abuse, otherwise than by good-humoured retorts, and by prosecuting his observations with renewed assiduity and zeal.

The use of single lenses in aid of the sight had been long known. Spectacles were in common use at the beginning of the fourteenth century, and there are several hints, more or less obscure, in many early writers, of the effects which might be expected from a combination of glasses; but it does not appear with certainty that any of these authors had attempted to reduce their ideas to practice. After the discovery of the telescope, almost every country endeavoured to find in the writings of its early philosophers traces of the knowledge of such an instrument, but in general with success very inadequate to the zeal of their national prepossessions. There are two authors especially to whom the attention of Kepler and others was turned, immediately upon the promulgation of the discovery, as containing the germ of it in their works. These are Baptista Porta, and Gerolamo Fracastoro. We have already had occasion to quote the Homocentrica of Fracastoro, who died in 1553; the following expressions, though they seem to refer to actual experiment, yet fall short of the meaning with which it has been attempted to invest them. After explaining and commenting on some phenomena of refraction through different media, to which he was led by the necessity of reconciling his theory with the variable magnitudes of the planets, he goes on to say—"For which reason, those things which are seen at the bottom of water, appear greater than those which are at the top; and if any one look through two eyeglasses, *one placed upon the other*, he will see every thing much larger and nearer."* It should seem that this passage (as Delambre has already remarked) rather refers to the close application of one glass upon another, and it may fairly be doubted whether any thing analogous to the composition of the telescope was in the writer's thoughts. Baptista Porta writes on the same subject more fully; —" Concave lenses show distant objects most clearly, convex those which are nearer, whence they may be used to assist the sight. With a concave glass distant objects will be seen, small, but distinct; with a convex one those near at hand, larger, but confused; *if you*

---

* Nova Sterometria Doliorum—Lincii, 1615.
† Mecanique Analytique.
‡ Histoire des Mathématiques, tom. ii.

* "Per duo specilla ocularia si quis perspiciat, altero alteri superposito, majora multo et propinquiora videbit omnia."—Fracast. Homocentrica, § 2, c. 8.

know rightly how to combine one of each sort, you will see both far and near objects larger and clearer."* These words show, if Porta really was then unacquainted with the telescope, how close it is possible to pass by an invention without lighting on it, for of precisely such a combination of a convex and concave lens, fitted to the ends of an organ pipe by way of tube, did the whole of Galileo's telescope consist. If Porta had stopped here he might more securely have enjoyed the reputation of the invention, but he then professes to describe the construction of his instrument, which has no relation whatever to his previous remarks. " I shall now endeavour to show in what manner, we may contrive to recognize our friends at the distance of several miles, and how those of weak sight may read the most minute letters from a distance. It is an invention of great utility, and grounded on optical principles, nor is it at all difficult of execution; but it must be so divulged as not to be understood by the vulgar, and yet be clear to the sharpsighted." The description which follows seems far enough removed from the apprehended danger of being too clear, and indeed every writer who has hitherto quoted it has merely given the passage in its original Latin, apparently despairing of an intelligible translation. With some alterations in the punctuation, which appear necessary to bring it into any grammatical construction,† it may be supposed to bear something like the following meaning:—" Let a view be contrived in the centre of a mirror, where it is most effective. All the solar rays are exceedingly dispersed, and do not in the least come together (in the true centre); but there is a concourse of all the rays in the central part of the said mirror, half way towards the other centre, where the cross diameters meet. This view is contrived in the following manner. A concave cylindrical mirror placed directly in front, but with its axis inclined, must be adapted to that focus: and let obtuse angled or right angled triangles be cut out with two cross lines on each side drawn from the centre, and a glass (*specillum*) will be completed fit for the purposes we mentioned." If it were not for the word " *specillum*," which, in the passage immediately preceding this, Porta* contrasts with " *speculum*," and which he afterwards explains to mean a glass lens, it would be very clear that the foregoing passage (supposing it to have any meaning) must be referred to a reflecting telescope, and it is a little singular that while this obscure passage has attracted universal attention, no one, so far as we are aware, has taken any notice of the following unequivocal description of the principal part of Newton's construction of the same instrument. It is in the 5th chapter of the 17th book, where Porta explains by what device exceedingly minute letters may be read without difficulty. " Place a concave mirror so that the back of it may lie against your breast; opposite to it, and within the burning point, place the writing; 'put a plane mirror behind it, that may be under your eyes. Then the images of the letters which are in the concave mirror, and which the concave has magnified, will be reflected in the plane mirror, so that you may read without difficulty."

We have not been able to meet with the Italian translation of Porta's Natural Magic, which was published in 1611, under his own superintendence; but the English translator of 1658 would probably have known if any intelligible interpretation were there given of the mysterious passage above quoted, and his translation is so devoid of meaning as strongly to militate against this idea. Porta, indeed, claimed the invention as his own, and is believed to have hastened his death, (which happened in 1615, he being then 80 years old,) by the fatigue of composing a Treatise on the Telescope, in which he had promised to exhaust the subject. We do not know whether this is the same work which was published after his death by Stelliola,† but which contains no allusion to Porta's claim, and possibly Stelliola may have thought it most for his friend's reputation to suppress it. Schott‡ says, a friend of his had

---

*: *(Si utrumque recte componere noveris, et longinqua et proxima majora et clara videbis.—Mag. Nat. lib. 17.

† The passage in the original, which is printed alike in the editions of 1598, 1607, 1619, and 1650, is as follows: Visus constituatur centro valentissimus speculi; ubi fiet, et Valentissimè universales solares radii disperguntur, et coeunt minimè, sed centro prædicti speculi in illius medio, ubi diametri transversales, omnium ibi concursus. Constituitur hoc modo speculum concavum columnare æquidistantibus lateribus, sed lateri uno obliquo sectionibus,illis accomodetur, trianguli vero obtusianguli, vel orthogonii secentur, hinc inde duobus transversalibus lineis, excentro eductis. Et confectum erit specillum, ad id, quod diximus, utile.

*. Diximus de Ptolemæi *speculo*,sive *specillo* potius, quo per sexcentena millia pervenientes naves conspiciebat. † Il Telescopio, 1627.
‡ Magia Naturæ et Artis Herbipoli, 1657.

seen Porta's book in manuscript, and that it did at that time contain the assertion of Porta's title to the invention. After all it is not improbable that he may have derived his notions of magnifying distant objects from our celebrated countryman Roger Bacon, who died about the year 1300. He has been supposed, not without good grounds, to have been one of the first who recognised the use of single lenses in producing distinct vision, and he has some expressions with respect to their combination which promise effects analogous to those held out by Porta. In "The Admirable Force of Art and Nature," he says, "Physical figurations are far more strange, for in such manner may we frame perspects and looking-glasses that one thing shall appear to be many, as one man shall seeme a whole armie; and divers sunnes and moones, yea, as many as we please, shall appeare at one time, &c. And so may the perspects be framed, that things most farre off may seeme most nigh unto us, and clean contrarie, soe that we may reade very small letters an incredible distance from us, and behold things how little soever they be, and make stars to appeare wheresoever we will, &c. And, besides all these, we may so frame perspects that any man entering into a house he shall indeed see gold, and silver, and precious stones, and what else he will, but when he maketh haste to the place he shall find just nothing." It seems plain, that the author is here speaking solely of mirrors, and we must not too hastily draw the conclusion, because in the first and last of these assertions he is, to a certain extent, borne out by facts, that he therefore was in possession of a method of accomplishing the middle problem also. In the previous chapter, he gives a long list of notable things, (much in the style of the Marquis of Worcester's Century of Iuventions) which if we can really persuade ourselves that he was capable of accomplishing, we must allow the present age to be still immeasurably inferior to him in science.

Thomas Digges, in the preface to his Pantometria, (published in 1591) declares, " My father, by his continuall painfull practises, assisted with demonstrations mathematicall, was able, and sundry times hath by proportionall glasses, duely situate in convenient angles, not only discouered things farre off, read letters, numbered peeces of money, with the verye coyne and superscription thereof, cast by some of his freends of purpose, upon downes in open fields; but also, seuen miles off, declared what hath beene doone at that instant in priuate places. He hath also sundrie times, by the sunne beames, fired powder and dischargde ordnance halfe a mile and more distante; which things I am the boulder to report, for that there are yet living diverse (of these his dooings) occulati testes, (eye witnesses) and many other matters farre more strange and rare, which I omit as impertinent to this place."

We find another pretender to the honour of the discovery of the telescope in the celebrated Antonio de Dominis, Archbishop of Spalatro, famous in the annals of optics for being one of the first to explain the theory of the rainbow. Montucla, following P. Boscovich, has scarcely done justice to De Dominis, whom he treats as a mere pretender and ignorant person. The indisposition of Boscovich towards him is sufficiently accounted for by the circumstance of his being a Catholic prelate who had embraced the cause of Protestantism. His nominal reconciliation with the Church of Rome would probably not have saved him from the stake, had not a natural death released him when imprisoned on that account at Rome. Judgment was pronounced upon him notwithstanding, and his body and books were publicly burnt in the Campo de' Fiori, in 1624. His treatise, De Radiis, (which is very rarely to be met with) was published by Bartolo after the acknowledged invention of the telescope by Galileo; but Bartolo tells us, in the preface, that the manuscript was communicated to him from a collection of papers written 20 years before, on his inquiring the Archbishop's opinion with respect to the newly discovered instrument, and that he got leave to publish it, " with the addition of one or two chapters." The treatise contains a complete description of a telescope, which, however, is professed merely to be an improvement on spectacles, and if the author's intention had been to interpolate an afterwritten account, in order to secure to himself the undeserved honour of the invention, it seems improbable that he would have suffered an acknowledgment of additions, previous to publication, to be inserted in the preface. Besides, the whole tone of the work is that of a candid and truth-seeking philosopher, very far indeed removed from being, as Montucla calls him, conspicuous for ignorance even among the ignorant men of his age. He gives a drawing of a convex and concave lens, and traces the passage of the rays through them; to which he subjoins, that he has not satisfied himself with any determination of the precise distance to which the glasses should be separated, according to their convexity and concavity, but recommends the proper distance to be found by actual experiment, and tells us, that the effect of the instrument will be to prevent the confusion arising from the interference of the direct and refracted rays, and to magnify the object by increasing the visible angle under which it is viewed. These, among the many claimants, are certainly the authors who approached the most nearly to the discovery: and the reader may judge, from the passages cited, whether the knowledge of the telescope can with probability be referred to a period earlier than the commencement of the 17th century. At all events, we can find no earlier trace of its being applied to any practical use; the knowlege, if it existed, remained speculative and barren.

In 1609, Galileo, then being on a visit to a friend at Venice, heard a rumour of the recent invention, by a Dutch spectacle-maker, of an instrument which was said to represent distant objects nearer than they usually appeared. According to his own account, this general rumour, which was confirmed to him by letters from Paris, was all that he learned on the subject; and returning to Padua, he immediately applied himself to consider the means by which such an effect could be produced. Fuccarius, in an abusive letter which he wrote on the subject, asserts that one of the Dutch telescopes had been at that time actually brought to Venice, and that he (Fuccarius) had seen it; which, even if true, is perfectly consistent with Galileo's statement; and in fact the question, whether or not Galileo saw the original instrument, becomes important only from his expressly asserting the contrary, and professing to give the train of reasoning by which he discovered its principle; so that any insinuation that he had actually seen the Dutch glass, becomes a direct impeachment of his veracity. It is certain, from the following extract of a letter from Lorenzo Pignoria to Paolo Gualdo, that one at least of the Dutch glasses had been sent to Italy. It is dated Padua, 31st August, 1609.*
" We have no news, except the return of His Serene Highness, and the reelection of the lecturers, among whom Sign. Galileo has contrived to get 1000 florins for life; and it is said to be on account of an eyeglass, *like the one which was sent from Flanders to Cardinal Borghese.* We have seen some here, and truly they succeed well."

It is allowed by every one that the Dutchman, or rather Zealander, made his discovery by mere accident, which greatly derogates from any honour attached to it; but even this diminished degree of credit has been fiercely disputed. According to one account, which appears consistent and probable, it had been made for sometime before its importance was in the slightest degree understood or appreciated, but was set up in the optician's shop as a curious philosophical toy, showing a large and inverted image of a weathercock, towards which it was directed. The Marquis Spinola, chancing to see it, was struck with the phenomenon, purchased the instrument, and presented it either to the Archduke Albert of Austria, or to Prince Maurice of Nassau, whose name appears in every version of the story, and who first entertained the idea of employing it in military reconnoissances.

Zacharias Jansen, and Henry Lipperhey, two spectacle-makers, living close to each other, near the church of Middleburg, have both had strenuous supporters of their title to the invention. A third pretender appeared afterwards in the person of James Metius of Alkmaer, who is mentioned by Huyghens and Des Cartes, but his claims rest upon no authority whatever comparable to that which supports the other two. About half a century afterwards, Borelli was at the pains to collect and publish a number of letters and depositions which he procured, as well on one side as on the other.† It seems that the truth lies between them, and that one, probably Jansen, was the inventor of the *microscope*, which application of the principle was unquestionably of an earlier date, perhaps as far back as 1590. Jansen gave one of his microscopes to the Archduke, who gave it to Cornelius Drebbel, a salaried mathematician at the court of our James the first, where William Borelli (not the author above

---

* Lettère d'Uomini illustri. Venezia, 1744.
† Borelli, De Vero Telescopii inventore, 1655.

mentioned) saw it many years afterwards, when ambassador from the United Provinces to England, and got from Drebbel this account of the quarter whence it came. Lipperhey afterwards, in 1609, accidentally hit upon the *telescope*, and on the fame of this discovery it would not be difficult for Jansen, already in possession of an instrument so much resembling it, to perceive the slight difference between them, and to construct a telescope independently of Lipperhey, so that each, with some show of reason, might claim the priority of the invention. A notion of this kind reconciles the testimony of many conflicting witnesses on the subject, some of whom do not seem to distinguish very accurately whether the telescope or microscope is the instrument to which their evidence refers. Borelli arrives at the conclusion, that Jansen was the inventor; but not satisfied with this, he endeavours, with a glaring partiality which makes his former determination suspicious, to secure for him and his son the more solid reputation of having anticipated Galileo in the useful employment of the invention. He has however inserted in his collections a letter from John the son of Zacharias, in which John, omitting all mention of his father, speaks of his own observation of the satellites of Jupiter, evidently seeking to insinuate that they were earlier than Galileo's; and in this sense the letter has since been quoted,* although it appears from John's own deposition, preserved in the same collection, that at the time of their discovery he could not have been more than six years old. An oversight of this sort throws doubt on the whole of the pretended observations, and indeed the letter has much the air of being the production of a person imperfectly informed on the subject on which he writes, and probably was compiled to suit Borelli's purposes, which were to make Galileo's share in the invention appear as small as possible.

Galileo himself gives a very intelligible account of the process of reasoning, by which he detected the secret.— "I argued in the following manner. The contrivance consists either of one glass or of more—one is not sufficient, since it must be either convex, concave, or plane; the last does not produce any sensible alteration in objects, the concave diminishes them: it is true that the convex magnifies, but it renders them confused and indistinct; consequently, one glass is insufficient to produce the desired effect. Proceeding to consider two glasses, and bearing in mind that the plane glass causes no change, I determined that the instrument could not consist of the combination of a plane glass with either of the other two. I therefore applied myself to make experiments on combinations of the 'two other kinds, and thus obtained that of which I was in search." It has been urged against Galileo that, if he really invented the telescope on theoretical principles, the same theory ought at once to have conducted him to a more perfect instrument than that which he at first constructed;* but it is plain, from this statement, that he does not profess to have theorized beyond the determination of the species of glass which he should employ in his experiments, and the rest of his operations he avows to have been purely empirical. Besides, we must take into account the difficulty of grinding the glasses, particularly when fit tools were yet to be made, and something must be attributed to Galileo's eagerness to bring his results to the test of actual experiment, without waiting for that improvement which a longer delay might and did suggest. Galileo's language bears a resemblance to the first passage which we quoted from Baptista Porta, sufficiently close to make it not improbable that he might be assisted in his inquiries by some recollection of it, and the same passage seems, in like manner, to have recurred to the mind of Kepler, as soon as he heard of the invention. Galileo's telescope consisted of a plano-convex and plano-concave lens, the latter nearest the eye, distant from each other by the difference of their focal lengths, being, in principle, exactly the same with the modern opera-glass. He seems to have thought that the Dutch glass was the same, but this could not be the case, if the above quoted particular of the *inverted* weathercock, which belongs to most traditions of the story, be correct; because it is the peculiarity of this kind of telescope not to invert objects, and we should be thus furnished with a demonstrative proof of the falsehood of 'Fuccarius's insinuation: in that case the Dutch glass must have been similar to what was afterwards called the astronomical telescope, consisting of two

---

* Encyclopædia Britannica, Art, TELESCOPE,   • Ibid,

convex glasses distant from each other by the sum of their focal lengths. This supposition is not controverted by the fact, that this sort of telescope was never employed by astronomers till long afterwards; for the fame of Galileo's observations, and the superior excellence of the instruments constructed under his superintendence, induced every one in the first instance to imitate his constructions as closely as possible. The astronomical telescope was however eventually found to possess superior advantages over that which Galileo imagined, and it is on this latter principle that all modern refracting telescopes are constructed; the inversion being counteracted in those which are intended for terrestrial observations, by the introduction of a second pair of similar glasses, which restore the inverted image to its original position. For further details on the improvements which have been subsequently introduced, and on the reflecting telescope, which was not brought into use till the latter part of the century, the reader is referred to the Treatise on OPTICAL INSTRUMENTS.

Galileo, about the same time, constructed microscopes on the same principle, for we find that, in 1612, he presented one to Sigismund, King of Poland; but his attention being principally devoted to the employment and perfection of his telescope, the microscope remained a long time imperfect in his hands: twelve years later, in 1624, he wrote to P. Federigo Cesi, that he had delayed to send the microscope, the use of which he there describes, because he had only just brought it to perfection, having experienced some difficulty in working the glasses. Schott tells an amusing story, in his "Magic of Nature," of a Bavarian philosopher, who, travelling in the Tyrol with one of the newly invented microscopes about him, was taken ill on the road and died. The authorities of the village took possession of his baggage, and were proceeding to perform the last duties to his body, when, on examining the little glass instrument in his pocket, which chanced to contain a flea, they were struck with the greatest astonishment and terror, and the poor Bavarian, condemned by acclamation as a sorcerer who was in the habit of using a portable familiar, was declared unworthy of Christian burial. Fortunately for his character, some bold sceptic ventured to open the instrument, and discovered the true nature of the imprisoned fiend.

As soon as Galileo's first telescope was completed, he returned with it to Venice, and the extraordinary sensation which it excited tends also strongly to refute Fuccarius's assertion that the Dutch glass was already known there. During more than a month Galileo's whole time was employed in exhibiting his instrument to the principal inhabitants of Venice, who thronged to his house to satisfy themselves of the truth of the wonderful stories in circulation; and at the end of that time the Doge, Leonardo Donati, caused it to be intimated to him that such a present would not be deemed unacceptable by the senate. Galileo took the hint, and his complaisance was rewarded by a mandate confirming him for life in his professorship at Padua, at the same time doubling his yearly salary, which was thus made to amount to 1000 florins.

It was long before the phrenzy of public curiosity abated. Sirturi describes a ludicrous violence which was done to himself, when, with the first telescope which he had succeeded in making, he went up into the tower of St. Mark, at Venice, in the vain hope of being there entirely unmolested. Unluckily he was seen by some idlers in the street: a crowd soon collected round him, who insisted on taking possession of his instrument, and, handing it one to the other, detained him there for several hours till their curiosity was satiated, when he was allowed to return home. Hearing them also inquire eagerly at what inn he lodged, he thought it better to quit Venice early the next morning, and prosecute his observations in a less inquisitive neighbourhood.* Instruments of an inferior description were soon manufactured, and vended everywhere as philosophical playthings, much in the way in which, in our own time, the kaleidoscope spread over Europe as fast as travellers could carry them. But the fabrication of a better sort was long confined, almost solely, to Galileo and those whom he immediately instructed; and so late as the year 1637, we find Gaertner, or as he chose to call himself, Hortensius, assuring Galileo that none could be met with in Holland sufficiently good to show Jupiter's disc well defined; and in 1634 Gassendi begs for a telescope from Galileo, informing

---

* Telescopium, Venetiis, 1619.

him that he was unable to procure a good one either in Venice, Paris, or Amsterdam.

The instrument, on its first invention, was generally known by the names of Galileo's tube, the perspective, the double eye-glass: the names of telescope and microscope were suggested by Demisiano, as we are told by Lagalla in his treatise on the Moon.*

CHAPTER VII.

Discovery of Jupiter's satellites—Kepler —Sizzi — Astrologers — Mæstlin— Horky—Mayer.

As soon as Galileo had provided himself with a second instrument, he began a careful examination of the heavenly bodies, and a series of splendid discoveries soon rewarded his diligence. After considering the beautiful appearances which the varied surface of the moon presented to this new instrument, he turned his telescope towards Jupiter, and his attention was soon arrested by the singular position of three small stars, near the body of that planet, which appeared almost in a straight line with it, and in the direction of the ecliptic. The following evening he was surprised to find that two of the three which had been to the eastward of the planet, now appeared on the contrary side, which he could not reconcile with the apparent motion of Jupiter among the fixed stars, as given by the tables. Observing these night after night, he could not fail to remark that they changed their relative positions. A fourth also appeared, and in a short time he could no longer refuse to believe that these small stars were four moons, revolving round Jupiter in the same manner in which our earth is accompanied by its single attendant. In honour of his patron Cosmo, he named them the Medicæan stars. As they are now hardly known by this appellation, his doubts, whether he should call them Medicæan, after Cosmo's family, or Cosmical, from his individual name, are become of less interest.

An extract from a letter which Galileo received on this occasion from the court of France, will serve to show how highly the honour of giving a name to these new planets was at that time appreciated, and also how much was expected from Galileo's first success in examining the heavens. "The second request, but the most pressing one which I can make to you, is, that you should determine, if you discover any other fine star, to call it by the name of the great star of France, as well as the most brilliant of all the earth; and, if it seems fit to you, call it rather by his proper name of Henri, than by the family name of Bourbon: thus you will have an opportunity of doing a thing just and due and proper in itself, and at the same time will render yourself and your family rich and powerful for ever." The writer then proceeds to enumerate the different claims of Henri IV. to this honour, not forgetting that he married into the family of the Medici, &c.

The result of these observations was given to the world, in an Essay which Galileo entitled *Nuncius Sidereus*, or the Intelligencer of the Stars; and it is difficult to describe the extraordinary sensation which its publication produced. Many doubted, many positively refused to believe, so novel an announcement; all were struck with the greatest astonishment, according to their respective opinions, either at the new view of the universe thus offered to them, or at the daring audacity of Galileo in inventing such fables. We shall proceed to extract a few passages from contemporary writers relative to this book, and the discoveries announced in it.

Kepler deserves precedence, both from his own celebrity, and from the lively and characteristic account which he gives of his first receiving the intelligence:—" I was sitting idle at home, thinking of you, most excellent Galileo, and your letters, when the news was brought me of the discovery of four planets by the help of the double eye-glass. Wachenfels stopped his carriage at my door to tell me, when such a fit of wonder seized me at a report which seemed so very absurd, and I was thrown into such agitation at seeing an old dispute between us decided in this way, that between his joy, my colouring, and the laughter of both, confounded as we were by such a novelty, we were hardly capable, he of speaking, or I of listening. My amazement was increased by the assertion of Wachenfels, that those who sent this news from Galileo were celebrated men, far removed by their learning, weight, and character, above vulgar folly; that the book was actually in the press, and would be published immediately. On our separating, the authority of Galileo had the greatest influence on

---

* De phænomenis in orbe Lunæ. Venetiis, 1612.

me, earned by the accuracy of his judgment, and excellence of his understanding; so I immediately fell to thinking how there could be any addition to the number of the planets without overturning my Mysterium Cosmographicum, published thirteen years ago, according to which Euclid's five regular solids do not allow more than six planets round the sun."

This was one of the many wild notions of Kepler's fanciful brain, among which he was lucky enough at length to hit upon the real and principal laws of the planetary motions. His theory may be briefly given in his own words:—" The orbit of the earth is the measure of the rest. About it circumscribe a dodecahedron. The sphere including this will be that of Mars. About Mars' orbit describe a tetrahedron: the sphere containing this will be Jupiter's orbit. Round Jupiter's describe a cube: the sphere including this will be Saturn's. Within the earth's orbit inscribe an icosahedron: the sphere inscribed in it will be Venus's orbit. In Venus inscribe an octahedron: the sphere inscribed in it will be Mercury's. You have now the reason of the number of the planets:" for as there are no more than the five regular solids here enumerated, Kepler conceived this to be a satisfactory reason why there could be neither more nor less than six planets. His letter continues :—" I am so far from disbelieving the existence of the four circumjovial planets, that I long for a telescope to anticipate you, if possible, in discovering two round Mars, (as the proportion seems to me to require,) six or eight round Saturn, and perhaps one each round Mercury and Venus."

The reader has here an opportunity of verifying Galileo's observation, that Kepler's method of philosophizing differed widely from his own. The proper line is certainly difficult to hit between the mere theorist and the mere observer. It is not difficult at once to condemn the former, and yet the latter will deprive himself of an important, and often indispensable assistance, if he neglect from time to time to consolidate his observations, and thence to conjecture the course of future observation most likely to reward his assiduity. This cannot be more forcibly expressed than in the words of Leonardo da Vinci :* " Theory is the general, experiments are the soldiers. The interpreter of the works of nature is experiment; that is never

* Venturi, Essai sur les ouvrages de Leo. da Vinci.

wrong; it is our judgment which is sometimes deceived, because we are expecting results which experiment refuses to give. We must consult experiment, and vary the circumstances, till we have deduced general rules, for it alone can furnish us with them. But you will ask, what is the use of these general rules? I answer, that they direct us in our inquiries into nature and the operations of art. They keep us from deceiving ourselves and others, by promising ourselves results which we can never obtain."

In the instance before us, it is well known that, adopting some of the opinions of Bruno and Brutti, Galileo, even before he had seen the satellites of Jupiter, had allowed the possibility of the discovery of new planets; and we can scarcely suppose that they had weakened his belief in the probability of further success, or discouraged him from examining the other heavenly bodies. Kepler on the contrary had taken the opposite side of the argument; but no sooner was the fallacy of his first position undeniably demonstrated, than, passing at once from one extreme to the other, he framed an unsupported theory to account for the number of satellites which were round Jupiter, and for those which he expected to meet with elsewhere. Kepler has been styled the legislator of the skies; his laws were promulgated rather too arbitrarily, and they often failed, as all laws must do which are not drawn from a careful observation of the nature of those who are to be governed by them. Astronomers have reason to be grateful for the theorems which he was the first to establish; but so far as regards the progress of the science of inductive reasoning, it is perhaps to be regretted, that the seventeen years which he wasted in random and unconnected guesses should have been finally rewarded, by discoveries splendid enough to shed deceitful lustre upon the method by which he arrived at them.

Galileo himself clearly perceived the fallacious nature of these speculations on numbers and proportions, and has expressed his sentiments concerning them very unequivocally. " How great and common an error appears to me the mistake of those who persist in making their knowledge and apprehension the measure of the apprehension and knowledge of God; as if that alone were perfect, which they understand to be so. But I, on the contrary, observe that

Nature has other scales of perfection, which we cannot comprehend, and rather seem disposed to class among imperfections. For instance, among the relations of different numbers, those appear to us most perfect which exist between numbers nearly related to each other; as the double, the triple, the proportion of three to two, &c.; those appear less perfect which exist between numbers remote from, and prime to each other; as 11 to 7, 17 to 13, 53 to 37, &c.; and most imperfect of all do those appear which exist between incommensurable quantities, which by us are nameless and inexplicable. Consequently, if the task had been given to a man, of establishing and ordering the rapid motions of the heavenly bodies, according to his notions of perfect proportions, I doubt not that he would have arranged them according to the former rational proportions; but, on the contrary, God, with no regard to our imaginary symmetries, has ordered them in proportions not only incommeasurable and irrational, but altogether inappreciable by our intellect. A man ignorant of geometry may perhaps lament, that the circumference of a circle does not happen to be exactly three times the diameter, or in some other assignable proportion to it, rather than such that we have not yet been able to explain what the ratio between them is; but one, who has more understanding will know that if they were other than they are, thousands of admirable conclusions would have been lost, and that none of the other properties of the circle would have been true: the surface of the sphere would not be quadruple of a great circle, nor the cylinder be to the sphere as three to two: in short, no part of geometry would be true, and as it now is. If one of our most celebrated architects had had to distribute this vast multitude of fixed stars through the great vault of heaven, I believe he would have disposed them with beautiful arrangements of squares, hexagons, and octagons; he would have dispersed the larger ones among the middle sized and the less, so as to correspond exactly with each other; and then he would think he had contrived admirable proportions: but God, on the contrary, has shaken them out from His hand as if by chance, and we, forsooth, must think that He has scattered them up yonder without any regularity, symmetry, and elegance."

It is worth remarking that the dangerous ideas of aptitude and congruence of numbers had taken such deep and general root, that long afterwards, when the reality of Jupiter's satellites was incontestably established, and Huyghens had discovered a similar satellite near Saturn, he was so rash as to declare his belief, (unwarned by the vast progress which astronomy had made in his own time,) that no more satellites would be discovered, since the one which he discovered near Saturn, with Jupiter's four, and our moon, made up the number six, exactly equal to the number of the principal planets. Every reader knows that this notion, so unworthy the genius of Huyghens, has been since exploded by the discovery both of new planets, and new satellites.

Francesco Sizzi, a Florentine astronomer, took the matter up in a somewhat different strain from Kepler.*— " There are seven windows given to animals in the domicile of the head, through which the air is admitted to the rest of the tabernacle of the body, to enlighten, to warm, and nourish it, which are the principal parts of the $\mu\iota\varkappa\rho o\varkappa o\sigma\mu o\varsigma$ (or little world); two nostrils, two eyes, two ears, and a mouth; so in the heavens, as in a $\mu\alpha\varkappa\rho o\varkappa o\sigma\mu o\varsigma$ (or great world), there are two favourable stars, two unpropitious, two luminaries, and Mercury alone undecided and indifferent. From which and many other similar phenomena of nature, such as the seven metals, &c., which it were tedious to enumerate, we gather that the number of planets is necessarily seven. Moreover, the satellites are invisible to the naked eye, and therefore can exercise no influence on the earth, and therefore would be useless, and therefore do not exist. Besides, as well the Jews and other ancient nations as modern Europeans have adopted the division of the week into seven days, and have named them from the seven planets: now if we increase the number of the planets this whole system falls to the ground." To these remarks Galileo calmly replied, that whatever their force might be, as a reason for believing beforehand that no more than seven planets would be discovered, they hardly seemed of sufficient weight to destroy the new ones when actually seen.

Others, again, took a more dogged line of opposition, without venturing into the subtle analogies and arguments of the philosopher just cited. They contented themselves, and satisfied others,

---

* Dianoia Astronomica, Venetiis, 1610.

with the simple assertion, that such things were not, and could not be, and the manner in which they maintained themselves in their incredulity was sufficiently ludicrous. "Oh, my dear Kepler,"* says Galileo, "how I wish that we could have one hearty laugh together. Here, at Padua, is the principal professor of philosophy, whom I have repeatedly and urgently requested to look at the moon and planets through my glass, which he pertinaciously refuses to do. Why are you not here? what shouts of laughter we should have at this glorious folly! and to hear the professor of philosophy at Pisa labouring before the grand duke with logical arguments, as if with magical incantations, to charm the new planets out of the sky."

Another opponent of Galileo deserves to be named, were it only for the singular impudence of the charge he ventures to bring against him. "We are not to think," says Christmann, in the Appendix to his *Nodus Gordius;* "that Jupiter has four satellites given him by nature, in order, by revolving round him, to immortalize the name of the Medici, who first had notice of the observation. These are the dreams of *idle men,* who love ludicrous ideas better than our laborious and industrious correction of the heavens.— Nature abhors so horrible a chaos, and to the truly wise such vanity is detestable.

Galileo was also urged by the astrologers to attribute some influence, according to their fantastic notions, to the satellites, and the account which he gives his friend Dini of his answer to one of this class is well worth extracting, as a specimen of his method of uniting sarcasm with serious expostulation; "I must," says he, "tell you what I said a few days back to one of those nativity-casters, who believe that God, when he created the heavens and the stars, had no thoughts beyond what they can themselves conceive, in order to free myself from his tedious importunity; for he protested, that unless I would declare to him the effect of the Medicæan planets, he would reject and deny them as needless and superfluous. I believe this set of men to be of Sizzi's opinion, that astronomers discovered the other seven planets, not by seeing them corporally in the skies, but only from their effects on earth,—much in the manner in which some houses are discovered to be haunted by evil spirits, not by seeing them, but from the extravagant pranks which are played there. I replied, that he ought to reconsider the hundred or thousand opinions which, in the course of his life, he might have given, and particularly to examine well the events which he had predicted with the help of Jupiter, and if he should find that all had succeeded conformably to his predictions, I bid him prophecy merrily on, according to his old and wonted rules; for I assured him that the new planets would not in any degree affect the things which are already past, and that in future he would not be a less fortunate conjuror than he had been: but if, on the contrary, he should find the events depending on Jupiter, in some trifling particulars not to have agreed with his dogmas and prognosticating aphorisms, he ought to set to work to find new tables for calculating the constitution of the four Jovial circulators at every bygone moment, and, perhaps, from the diversity of their aspects, he would be able, with accurate observations and multiplied conjunctions, to discover the alterations and variety of influences depending upon them; and I reminded him, that in ages past they had not acquired knowledge with little labour, at the expense of others, from written books, but that the first inventors acquired the most excellent knowledge of things natural and divine with study and contemplation of the vast book which nature holds ever open before those who have eyes in their forehead and in their brain; and that it was a more honourable and praiseworthy enterprize with their own watching, toil, and study, to discover something admirable and new among the infinite number which yet remain concealed in the darkest depths of philosophy, than to pass a listless and lazy existence, labouring only to darken the toilsome inventions of their neighbours, in order to excuse their own cowardice and inaptitude for reasoning, while they cry out that nothing can be added to the discoveries already made."

The extract given above from Kepler, is taken from an Essay, published with the later editions of the *Nuncius,* the object and spirit of which seem to have been greatly misunderstood, even by some of Kepler's intimate friends.— They considered it as a covert attack upon Galileo, and, accordingly, Maestlin thus writes to him:—"In your Essay

---

* Kepleri Epistolæ.

(which I have just received) you have plucked Galileo's feathers well; I mean, that you have shown him not to be the inventor of the telescope, not to have been the first who observed the irregularities of the moon's surface; not to have been the first discoverer of more worlds than the ancients were acquainted with, &c. One source of exultation was still left him, from the apprehension of which Martin Horky has now entirely delivered me." It is difficult to discover in what part of Kepler's book Maestlin found all this, for it is one continued encomium upon Galileo; insomuch that Kepler almost apologizes in the preface for what may seem his intemperate admiration of his friend. "Some might wish I had spoken in more moderate terms in praise of Galileo, in consideration of the distinguished men who are opposed to his opinions, but I have written nothing fulsome or insincere. I praise him, for myself; I leave other men's judgments free; and shall be ready to join in condemnation when some one wiser than myself shall, by sound reasoning, point out his errors." However, Maestlin was not the only one who misunderstood Kepler's intentions: the Martin Horky of whom he speaks, a young German, also signalized himself by a vain attack upon the book which he thought his patron Kepler condemned. He was then travelling in Italy, whence he wrote to Kepler his first undetermined thoughts about the new discoveries. "They are wonderful; they are stupendous; whether they are true or false I cannot tell."* He seems soon to have decided that most reputation was to be gained on the side of Galileo's opponents, and his letters accordingly became filled with the most rancorous abuse of him. At the same time, that the reader may appreciate Horky's own character, we shall quote a short sentence at the end of one of his letters, where he writes of a paltry piece of dishonesty with as great glee as if he had solved an ingenious and scientific problem. After mentioning his meeting Galileo at Bologna, and being indulged with a trial of his telescope, which, he says, "does wonders upon the earth, but represents celestial objects falsely;"† he concludes with the following honourable sentence:—" I must confide to you a theft which I committed. I contrived to take a mould of the glass in wax, without the knowledge of any one, and, when I get home, I trust to make a telescope even better than Galileo's own."

Horky having declared to Kepler, "I will never concede his four new planets to that Italian from Padua though I die for it," followed up this declaration by publishing a book against Galileo, which is the one alluded to by Maestlin, as having destroyed the little credit which, according to his view, Kepler's publication had left him. This book professes to contain the examination of four principal questions touching the alleged planets; 1st, Whether they exist? 2nd, What they are? 3rd, What they are like? 4th, Why they are? The first question is soon disposed of, by Horky's declaring positively that he has examined the heavens with Galileo's own glass, and that no such thing as a satellite about Jupiter exists. To the second, he declares solemnly, that he does not more surely know that he has a soul in his body, than that reflected rays are the sole cause of Galileo's erroneous observations. In regard to the third question, he says, that these planets are like the smallest fly compared to an elephant; and, finally, concludes on the fourth, that the only use of them is to gratify Galileo's "thirst of gold," and to afford himself a subject of discussion.*

Galileo did not condescend to notice this impertinent folly; it was answered by Roffini, a pupil of Magini, and by a young Scotchman of the name of Wedderburn, then a student at Padua, and afterwards a physician at the Court of Vienna. In the latter reply we find it mentioned, that Galileo was also using his telescope for the examination of insects,

* Kepleri Epistolæ.
† It may seem extraordinary that any one could support an argument by this partial disbelief in the instrument, which was allowed on all hands to represent terrestrial objects correctly. A similar instance of obstinacy, in an almost identical case though in a more unpretending station, once came under the writer's own observation. A farmer in Cambridgeshire, who had acquired some confused notions of the use of the quadrant, consulted him on a new method of determining the distances and magnitudes of the sun and moon, which he declared were far different from the quantities usually assigned to them. After a little conversation, the root of his error, certainly sufficiently gross, appeared to be that he had confounded the angular measure of a degree, with $69\frac{1}{2}$ miles, the linear measure of a degree on the earth's surface. As a short way of showing his mistake, he was desired to determine, in the same manner, the height of his barn which stood about 30 yards distant; he lifted the quadrant to his eye, but perceiving, probably, the monstrous size to which his principles were forcing him, he said, "Oh, Sir, the quadrant's only true for the sky." He must have been an objector of this kind, who said to Galileo,—" Oh, Sir, the telescope's only true for the earth."
* Venturi.

&c.* Horky sent his performance triumphantly to Kepler, and, as he returned home before receiving an answer, he presented himself before his patron in the same misapprehension under which he had written, but the philosopher received him with a burst of indignation which rapidly undeceived him. The conclusion of the story is characteristic enough to be given in Kepler's own account of the matter to Galileo, in which, after venting his wrath against this " scum of a fellow," whose " obscurity had given him audacity," he says, that Horky begged so hard to be forgiven, that " I have taken him again into favour upon this preliminary condition, to which he has agreed:—that I am to shew him Jupiter's satellites, AND HE IS TO SEE THEM, and own that they are there."

In the same letter Kepler writes, that although he has himself perfect confidence in the truth of Galileo's assertions, yet he wishes he could furnish him with some corroborative testimonies, which Kepler could quote in arguing the point with others. This request produced the following reply, from which the reader will also learn the new change which had now taken place in Galileo's fortunes, the result of the correspondence with Florence, part of which we have already extracted.† " In the first place, I return you my thanks that you first, and almost alone, before the question had been sifted (such is your candour and the loftiness of your mind), put faith in my assertions. You tell me you have some telescopes, but not sufficiently good to magnify distant objects with clearness, and that you anxiously expect a sight of mine, which magnifies images more than a thousand times. It is mine no longer, for the Grand Duke of Tuscany has asked it of me, and intends to lay it up in his museum, among his most rare and precious curiosities, in eternal remembrance of the invention: I have made no other of equal excellence, for the mechanical labour is very great: I have, however, devised some instruments for figuring and polishing them which I am unwilling to construct here, as they could not conveniently be carried to Florence, where I shall in future reside. You ask, my dear Kepler, for other testimonies:—I produce, for one, the Grand Duke, who, after observing the Medicæan planets several times with me at Pisa during the last months, made me a present, at parting, worth more than a thousand florins, and has now invited me to attach myself to him with the annual salary of one thousand florins, and with the title of Philosopher and Principal Mathematician to His Highness; without the duties of any office to perform, but with the most complete leisure; so that I can complete my Treatises on Mechanics, on the Constitution of the Universe, and on Natural and Violent Local Motion, of which I have demonstrated geometrically many new and admirable phenomena. I produce, for another witness, myself, who, although already endowed in this college with the noble salary of one thousand florins, such as no professor of mathematics ever before received, and which I might securely enjoy during my life, even if these planets had deceived me and should disappear, yet quit this situation, and betake me where want and disgrace will be my punishment should I prove to have been mistaken."

It is difficult not to regret that Galileo should be thus called on to resign his best glasses, but it appears probable that on becoming more familiar with the Grand Duke, he ventured to suggest that this telescope would be more advantageously employed in his own hands, than pompously laid up in a museum; for in 1637 we find him saying, in answer to a request from his friend Micanzio to send him a telescope—" I am sorry that I cannot oblige you with the glasses for your friend, but I am no longer capable of making them, and I have just parted with two tolerably good ones which I had, reserving only my old discoverer of celestial novelties which is already promised to the Grand Duke. Cosmo was dead in 1637, and it is his son Ferdinand who is here meant, who appears to have inherited his father's love of science. Galileo tells us, in the same letter, that Ferdinand had been amusing himself for some months with making object-glasses, and always carried one with him to work at wherever he went.

When forwarding this telescope to Cosmo in the first instance, Galileo adds, with a very natural feeling—" I send it to his highness unadorned and unpolished, as I made it for my own use, and beg that it may always be left in the same state; for none of the old parts ought to be displaced to make room for new ones, which will have had no share in the watchings and fatigues

---

* Quatuor probl. confut. per J. Wedderbornum, Scotobritannum. Patavii, 1610.
† See page 18.

of these observations." A telescope was in existence, though with the object glass broken, at the end of the last century, and probably still is in the Museum at Florence, which was shewn as the discoverer of Jupiter's satellites. Nelli, on whose authority this is mentioned, appears to question its genuineness. The first reflecting telescope, made with Newton's own hands, and scarcely possessing less interest than the first of Galileo's, is preserved in the library of the Royal Society.

By degrees the enemies of Galileo and of the new stars found it impossible to persevere in their disbelief, whether real or pretended, and at length seemed resolved to compensate for the sluggishness of their perception, by its acuteness when brought into action. Simon Mayer published his "Mundus Jovialis" in 1614, in which he claims to have been an original observer of the satellites, but, with an affectation of candour, allows that Galileo observed them probably about the same time. The earliest observation which he has recorded is dated 29th December, 1609, but, not to mention the total want of probability that Mayer would not have immediately published so interesting a discovery, it is to be observed, that, as he used the old style, this date of 29th December agrees with the 8th January, 1610, of the new style, which was the date of Galileo's second observation, and Galileo ventured to declare his opinion, that this pretended observation was in fact a plagiarism.

Scheiner counted five, Rheita nine, and other observers, with increasing contempt for Galileo's imperfect announcements, carried the number as high as twelve.* In imitation of Galileo's nomenclature, and to honour the sovereigns of the respective observers, these supposed additional satellites were dignified with the names of Vladislavian, Agrippine, Urbanoctavian, and Ferdinandotertian planets; but a very short time served to show it was as unsafe to exceed as to fall short of the number which Galileo had fixed upon, for Jupiter rapidly removed himself from the neighbourhood of the fixed stars, which gave rise to these pretended discoveries, carrying with him only his four original attendants, which continued in every part of his orbit to revolve regularly about him.

Perhaps we cannot better wind up this account of the discovery of Jupiter's satellites, and of the intense interest they have at all times inspired, than in the words of one who inherits a name worthy to be ranked with that of Galileo in the list of astronomical discoverers, and who takes his own place among the most accomplished mathematicians of the present times. " The discovery of these bodies was one of the first brilliant results of the invention of the telescope; one of the first great facts which opened the eyes of mankind to the system of the universe, which taught them the comparative insignificance of their own planet, and the superior vastness and nicer mechanism of those other bodies, which had before been distinguished from the stars only by their motion, and wherein none but the boldest thinkers had ventured to suspect a community of nature with our own globe. This discovery gave the holding turn to the opinions of mankind respecting the Copernican system; the analogy presented by these little bodies (little however only in comparison with the great central body about which they revolve) performing their beautiful revolutions in perfect harmony and order about it, being too strong to be resisted. This elegant system was watched with all the curiosity and interest the subject naturally inspired. The eclipses of the satellites speedily attracted attention, and the more when it was discerned, as it speedily was, by Galileo himself, that they afforded a ready method of determining the difference of longitudes of distant places on the earth's surface, by observations of the instants of their disappearances and reappearances, simultaneously made. Thus the first astronomical solution of the great problem of the longitude, the first mighty step which pointed out a connection between speculative astronomy and practical utility, and which, replacing the fast dissipating dreams of astrology by nobler visions, showed how the stars might really, and without fiction, be called arbiters of the destinies of empires, we owe to the satellites of Jupiter, those atoms imperceptible to the naked eye, and floating like motes in the beam of their primary—itself an atom, to our sight, noticed only by the careless vulgar as a large star, and by the philosophers of former ages as something moving among the stars, they knew not what, nor why: perhaps only to perplex the wise with fruitless conjectures, and harass the weak with fears as idle as their theories."*

---

* Sherburne's Sphere of Manilius. London, 1675.

* Herschel's Address to the Astronomical Society, 1827.

Chapter VIII.

Observations on the Moon—Nebulæ—Saturn—Venus—Mars.

There were other discoveries announced in Galileo's book of great and unprecedented importance, and which scarcely excited less discussion than the controverted Medicæan planets. His observations on the moon threw additional light on the constitution of the solar system, and cleared up the difficulties which encumbered the explanation of the varied appearance of her surface. The different theories current at that day, to account for these phenomena, are collected and described by Benedetti, and also with some liveliness, in a mythological poem, by Marini.* We are told, that, in the opinion of some, the dark shades on the moon's surface arise from the interposition of opaque bodies floating between her and the sun, which prevents his light from reaching those parts: others thought, that on account of her vicinity to the earth, she was partly tainted with the imperfection of our terrestrial and elementary nature, and was not of that entirely pure and refined substance of which the more remote heavens consist: a third party looked on her as a vast mirror, and maintained that the dark parts of her surface were the reflected images of our earthly forests and mountains.

Galileo's glass taught him to believe that the surface of this planet, far from being smooth and polished, as was generally taken for granted, really resembled our earth in its structure; he was able distinctly to trace on it the outlines of mountains and other inequalities, the summits of which reflected the rays of the sun before these reached the lower parts, and the sides of which, turned from his beams, lay buried in deep shadow. He recognised a distribution into something similar to continents of land, and oceans of water, which reflect the sun's light to us with greater or less vivacity, according to their constitution. These conclusions were utterly odious to the Aristotelians; they had formed a preconceived notion of what the moon ought to be, and they loathed the doctrines of Galileo, who took delight, as they said, in distorting and ruining the fairest works of nature. It was in vain he argued, as to the imaginary perfection of the spherical form, that, although the moon, or the earth, were it absolutely smooth, would indeed be a more perfect sphere than in its present rough state, yet touching the perfection of the earth, considered as a natural body calculated for a particular purpose, every one must see that absolute smoothness and sphericity would make it not only less perfeet, but as far from being perfect as possible. " What else," he demanded, " would it be but a vast unblessed desert, void of animals, of plants, of cities and of men; the abode of silence and inaction; senseless, lifeless, soulless, and stript of all those ornaments which make it now so various and so beautiful?"

He reasoned to no purpose with the slaves of the ancient schools: nothing could console them for the destruction of their smooth unalterable surface, and to such an absurd length was this hallucination carried, that one opponent of Galileo, Lodovico delle Colombe, constrained to allow the evidence of the sensible inequalities of the moon's surface, attempted to reconcile the old doctrine with the new observations, by asserting, that every part of the moon, which to the terrestrial observer appeared hollow and sunken, was in fact entirely and exactly filled up with a clear crystal substance, perfectly imperceptible by the senses, but which restored to the moon her accurately spherical and smooth surface. Galileo met the argument in the manner most fitting, according to one of Aristotle's own maxims, that " it is foolish to refute absurd opinions with too much curiosity." " Truly," says he, " the idea is admirable, its only fault is that it is neither demonstrated nor demonstrable; but I am perfectly ready to believe it, provided that, with equal courtesy, I may be allowed to raise upon your smooth surface, crystal mountains (which nobody can perceive) ten times higher than those which I have actually seen and measured." By threatening to proceed to such extremities, he seems to have scared the opposite party into moderation, for we do not find that the crystalline theory was persevered in.

In the same essay, Galileo also explained at some length the cause of that part of the moon being visible, which is unenlightened directly by the sun in her first and last quarter. Maestlin, and before him Leonardo da Vinci, had already declared this to arise from what may be called *earthshine*, or the reflec-

---

* Adone di Marini, Venetiis, 1623, Cant. x.

tion of the sun's light from the terrestrial globe, exactly similar to that which the moon affords us when we are similarly placed between her and the sun; but the notion had not been favourably received, because one of the arguments against the earth being a planet, revolving like the rest round the sun, was, that it did not shine like them, and was therefore of a different nature; and this argument, weak as it was in itself, the theory of terrestrial reflection completely overturned. The more popular opinions ascribed this feeble light, some to the fixed stars, some to Venus, some to the rays of the sun, penetrating and shining through the moon. Even the sagacious Benedetti adopted the notion of this light being caused by Venus, in the same sentence in which he explains the true reason of the faint light observed during a total eclipse of the moon, pointing out that it is occasioned by those rays of the sun, which reach the moon, after being bent round the sides of the earth by the action of our atmosphere.*

Galileo also announced the detection of innumerable stars, invisible to the unassisted sight; and those remarkable appearances in the heavens, generally called nebulæ, the most considerable of which is familiar to all under the name of the milky way, when examined by his instrument, were found to resolve themselves into a vast collection of minute stars, too closely congregated to produce a separate impression upon the unassisted eye.† Benedetti, who divined that the dark shades on the moon's surface arose from the constitution of those parts which suffered much of the light to pass into them, and consequently reflected a less portion of it, had maintained that the milky way was the result of the converse of the same phenomenon, and declared, in the language of his astronomy, that it was a part of the eighth orb, which did not, like the rest, allow the sun's light to traverse it freely, but reflected a small part feebly to our sight.

The Anti-Copernicans would probably have been well pleased, if by these eternally renewed discussions and disputes, they could have occupied Galileo's time sufficiently to detain his attention from his telescope and astronomical observations; but he knew too well where his real strength lay, and they had scarcely time to compound any thing like an argument against him and his theories, before they found him in possession of some new facts, which they were unprepared to meet, otherwise than by the never-failing resource of abuse and affected contempt. The year had not expired before Galileo had new intelligence to communicate of the highest importance. Perhaps he had been taught caution from the numerous piracies which had been committed upon his discoveries, and he first announced his new discoveries enigmatically, veiling their real import by transpositions of the letters in the words which described them, (a practice then common, and not disused even at a much later date,) and inviting all astronomers to declare, within a certain time, if they had noted any thing new in the heavens worthy of observation. The transposed letters which he published were—

"*Smaismrmilme poeta leumi bvne nugttaviras.*"

Kepler, in the true spirit of his riddling philosophy, endeavoured to decypher the meaning, and fancied he had succeeded when he formed a barbarous Latin verse,

"*Salve umbistineum geminatum Martia proles,*"

conceiving that the discovery, whatever it might be, related to the planet Mars, to which Kepler's attention had before been particularly directed. The reader, however, need not weary himself in seeking a translation of this solution, for at the request of the Emperor Rodolph, Galileo speedily sent to him the real reading—

Altissimum planetam tergeminum observavi;

that is, " I have observed that the most distant planet is triple," or, as he further explains the matter, " I have with great admiration observed that Saturn is not a single star, but three together, which as it were touch each other; they have no relative motion, and are constituted in this form oOo the middle being somewhat larger than the lateral ones. If we examine them with an eye-glass which magnifies the surface less than 1000 times, the three stars do not appear very distinctly, but Saturn has an oblong appearance, like the appearance of an olive, thus ⬭. Now I have discovered a court for Jupiter, and two servants for this old man, who aid his

---

* Speculat. Lib Venetiis, 1585, Epistolæ.
† This opinion, with respect to the milky way, had been held by some of the ancient astronomers. See Manilius. Lib. i. V. 753.
  " *Anne magis densâ stellarum turba coronâ*
  Contextit flammas, et crasso lumine candet,
  Et fulgore nitet collato clarior orbis."

steps and never quit his side." Galileo was, however, no match in this style of writing for Kepler, who disapproved his friend's metaphor, and, in his usual fanciful and amusing strain,—" I will not," said he, " make an old man of Saturn, nor slaves of his attendant globes, but rather let this tricorporate form be Geryon, so shall Galileo be Hercules, and the telescope his club; armed with which, he has conquered that distant planet, and dragged him from the remotest depths of nature, and exposed him to the view of all." Galileo's glass was not of sufficient power to shew him the real constitution of this extraordinary planet; it was reserved for Huyghens, about the year 1656, to declare to the world that these supposed attendant stars are in fact part of a ring which surrounds, and yet is completely distinct from the body of Saturn;* and the still more accurate observations of Herschel have ascertained that it consists of two concentric rings revolving round the planet, and separated from each other by a space which our most powerful telescopes scarcely enable us to measure.

Galileo's second statement concluded with the remark, that " in the other planets nothing new was to be observed;" but a month had scarcely elapsed, before he communicated to the world another enigma,

Hæc immatura à me jam frustra leguntur oy,

which, as he said, contained the announcement of a new phenomenon, in the highest degree important to the truth of the Copernican system. The interpretation of this is,

Cynthiæ figuras æmulatur mater amorum,

that is to say,—Venus rivals the appearances of the moon — for Venus being now arrived at that part of her orbit in which she is placed between the earth and the sun, and consequently, with only a part of her enlightened surface turned towards the earth, the telescope shewed her in a crescent form, like the moon in a similar position, and tracing her through the whole of her orbit round the sun, or at least so long as she was not invisible from his overpowering light, Galileo had the satisfaction of seeing the enlightened portion in each position assume the form appropriate to that hypothesis. It was with reason, therefore, that he laid stress on the importance of this observation, which also established another doctrine scarcely less obnoxious to the Anti-Copernicans, namely, that a new point of resemblance was here found between the earth and one of the principal planets; and as the reflection from the earth upon the moon had shewn it to be luminous like the planets when subjected to the rays of the sun, so this change of apparent figure demonstrated that one of the planets not near the earth, and therefore probably all, were in their own nature not luminous, and only reflected the sun's light which fell upon them; an inference, of which the probability was still farther increased a few years later by the observation of the transit of Mercury over the sun's disc.

It is curious that only twenty-five years before this discovery of the phases (or appearances) of Venus, a commentator of Aristotle, under the name of Lucilius Philalthæus, had advanced the doctrine that all the planets except the moon are luminous of themselves, and in proof of his assertion had urged, " that if the other planets and fixed stars received their light from the sun, they would, as they approached and receded from him, or as he approached and receded from them, assume the same phases as the moon, which, he adds, we have never yet observed."—He further remarks, " that Mercury and Venus would, in the supposed case of their being nearer the earth than the sun, eclipse it occasionally, just as eclipses are occasioned by the moon." Perhaps it is still more remarkable, that these very passages, in which the reasoning is so correct, though the facts are too hastily taken for granted, (the common error of that school,) are quoted by Benedetti, expressly to shew the ignorance and presumption of the author. Copernicus, whose want of instruments had prevented him from observing the horned appearance of Venus when between the earth and sun, had perceived how formidable an obstacle the non-appearance of this phenomenon presented to his system; he endeavoured, though unsatisfactorily, to account for it by supposing that the rays of the sun passed freely through the body of the planet, and Galileo takes occasion to praise him for not being deterred from

---

* Huyghens announced his discovery in this form : *aaaaaaacccccdeeeeeghiiiiiiiillllmmnn nnnnnnnoooopp q rrstttttuuuuu,* which he afterwards recomposed into the sentence. *Annulo cingitur, tenui, plano, nusquam cohærente, ad ecliptic-cam inclinato.* De Saturni Lunâ. Hagæ, 1656.

adopting the system, which, on the whole, appeared to agree best with the phenomena, by meeting with some which it did not enable him to explain. Milton, whose poem is filled with allusions to Galileo and his astronomy, has not suffered this beautiful phenomenon to pass unnoticed. After describing the creation of the Sun, he adds:—

Hither, as to their fountain, other stars
Repairing, in their golden urns draw light,
And hence the morning planet gilds her horns.*

Galileo also assured himself, at the same time, that the fixed stars did not receive their light from the sun. This he ascertained by comparing the vividness of their light, in all positions, with the feebleness of that of the distant planets, and by observing the different degrees of brightness with which all the planets shone at different distances from the sun. The more remote planets did not, of course, afford equal facilities with Venus for so decisive an observation; but Galileo thought he observed, that when Mars was in quadratures, (or in the quarters, the middle points of his path on either side,) his figure varied slightly from a perfect circle. Galileo concludes the letter, in which he announces these last observations to his pupil Castelli, with the following expressions, shewing how justly he estimated the opposition they encountered:—"You almost make me laugh by saying that these clear observations are sufficient to convince the most obstinate: it seems you have yet to learn that long ago the observations were enough to convince those who are capable of reasoning, and those who wish to learn the truth; but that to convince the obstinate, and those who care for nothing beyond the vain applause of the stupid and senseless vulgar, not even the testimony of the stars would suffice, were they to descend on earth to speak for themselves. Let us then endeavour to procure some knowledge for ourselves, and rest contented with this sole satisfaction; but of advancing in popular opinion, or gaining the assent of the book-philosophers, let us abandon both the hope and the desire."

CHAPTER IX.

Account of the Academia Lincea—Del Cimento—Royal Society.

GALILEO'S resignation of the mathematical professorship at Padua occasioned much dissatisfaction to all those who were connected with that university. Perhaps not fully appreciating his desire of returning to his native country, and the importance to him and to the scientific world in general, of the complete leisure which Cosmo secured to him at Florence, (for by the terms of his diploma he was not even required to reside at Pisa, nor to give any lectures, except on extraordinary occasions, to sovereign princes and other strangers of distinction,) the Venetians remembered only that they had offered him an honourable asylum when almost driven from Pisa; that they had increased his salary to four times the sum which any previous professor had enjoyed; and, finally, by an almost unprecedented decree, that they had but just secured him in his post during the remainder of his life. Many took such offence as to refuse to have any further communication with him; and Sagredo, a constant friend of Galileo, wrote him word that he had been threatened with a similar desertion unless he should concur in the same peremptory resolution, which threats, however, Sagredo, at the same time, intimates his intention of braving.

Early in the year 1611, Galileo made his first appearance in Rome, where he was received with marks of distinguished consideration, and where all ranks were eager to share the pleasure of contemplating the new discoveries. "Whether we consider cardinal, prince, or prelate, he found an honourable reception from them all, and had their palaces as open and free to him as the houses of his private friends."* Among other distinctions he was solicited to become a member of the newly-formed philosophical society, the once celebrated *Academia Lincea*, to which he readily assented. The founder of this society was Federigo Cesi, the Marchese di Monticelli, a young Roman nobleman, the devotion of whose time and fortune to the interests of science has not been by any means rewarded with a reputation commensurate with his deserts. If the energy of his mind had been less worthily employed than in fostering the cause of science and truth, and in extending the advantages of his birth and fortune to as many as were willing to co-operate with him, the name of Federigo Cesi might have appeared more prominently on the page of history. Cesi had scarcely completed

* B. vii. v. 364. Other passages may be examined in B. i. 296; iii. 565—590, 722—733; iv. 589; v. 261, 414; vii. 577; viii. 1—178.

* Salusbury, Math. Coll.

his 18th year, when, in 1603, he formed the plan of a philosophical society, which in the first instance consisted only of himself and three of his most intimate friends, Hecke, a Flemish physician, Stelluti, and Anastasio de Filiis. Cesi's father, the Duca d' Acquasparta, who was of an arbitrary and extravagant temper, considered such pursuits and associates as derogatory to his son's rank; he endeavoured to thwart the design by the most violent and unjustifiable proceedings, in consequence of which, Cesi in the beginning of 1605 privately quitted Rome, Hecke was obliged to leave Italy altogether from fear of the Inquisition, which was excited against him, and the academy was for a time virtually dissolved. The details of these transactions are foreign to the present narrative: it will be enough to mention that, in 1609, Cesi, who had never altogether abandoned his scheme, found the opposition decaying which he at first experienced, and with better success he renewed the plan which he had sketched six years before. A few extracts from the Regulations will serve to shew the spirit in which this distinguished society was conceived:—

"The Lyncean Society desires for its academicians, philosophers eager for real knowledge, who will give themselves to the study of nature, and especially to mathematics; at the same time it will not neglect the ornaments of elegant literature and philology, which like a graceful garment adorn the whole body of science.—In the pious love of wisdom, and to the praise of the most good and most high God, let the Lynceans give their minds, first to observation and reflection, and afterwards to writing and publishing.—It is not within the Lyncean plan to find leisure for recitations and declamatory assemblies; the meetings will neither be frequent nor full, and chiefly for transacting the necessary business of the society: but those who wish to enjoy such exercises will in no respect be hindered, provided they attend them as accessory studies, decently and quietly, and without making promises and professions of how much they are about to do. For there is ample philosophical employment for every one by himself, particularly if pains are taken in travelling and in the observation of natural phenomena, and in the book of nature which every one has at home, that is to say, the heavens and the earth; and enough may be learned from the habits of constant correspondence with each other, and alternate offices of counsel and assistance.—Let the first fruits of wisdom be love; and so let the Lynceans love each other as if united by the strictest ties, nor suffer any interruption of this sincere bond of love and faith, emanating from the source of virtue and philosophy. —Let them add to their names the title of Lyncean, which has been advisedly chosen as a warning and constant stimulus, especially when they write on any literary subject, also in their private letters to their associates, and in general when any work comes from them wisely and well performed.—The Lynceans will pass over in silence all political controversies and quarrels of every kind, and wordy disputes, especially gratuitous ones, which give occasion to deceit, unfriendliness, and hatred; like men who desire peace, and seek to preserve their studies free from molestation, and to avoid every sort of disturbance. And if any one by command of his superiors, or from some other necessity, is reduced to handle such matters, since they are foreign to physical and mathematical science, and consequently alien to the object of the Academy, let them be printed without the Lyncean name."*

The society which was eventually organized formed but a very trifling part of the comprehensive scheme which Cesi originally proposed to himself; it had been his wish to establish a scientific Order which should have corresponding lodges in the principal towns of Europe, and in other parts of the globe, each consisting of not more than five nor less than three members, besides an unlimited number of Academicians not restricted to any particular residence or regulations. The mortifications and difficulties to which he was subjected from his father's unprincipled behaviour, render it most extraordinary and admirable that he should have ventured to undertake even so much as he actually carried into execution. He promised to furnish to the members of his society such assistance as they might require in the prosecution of their respective researches, and also to defray the charges

* Perhaps it was to deprecate the hostility of the Jesuits that, at the close of these Regulations, the Lynceans are directed to address their prayers, among other Saints, especially to Ignatius Loyola, as to one who greatly favoured the interests of learning. Odescalchi, Memorie dell' Acad. de' Lincei, Roma. 1806.

of publishing such of their works as should be thought worthy of appearing with the common sanction. Such liberal offers were not likely to meet with an unfavourable reception: they were thankfully accepted by many well qualified to carry his design into execution, and Cesi was soon enabled formally to open his academy, the distinctive title of which he borrowed from the Lynx, with reference to the piercing sight which that animal has been supposed to possess. This quality seemed to him an appropriate emblem of those which he desired to find in his academicians, for the purpose of investigating the secrets of nature; and although, at the present day, the name may appear to border on the grotesque, it was conceived in the spirit of the age, and the fantastic names of the numberless societies which were rapidly formed in various parts of Italy far exceed whatever degree of quaintness may be thought to belong to the Lyncean name. The Inflamed — the Transformed — the Uneasy — the Humorists—the Fantastic—the Intricate—the Indolent—the Senseless—the Undeceived—the Valiant — the Ætherial Societies are selected from a vast number of similar institutions, the names of which, now almost their sole remains, are collected by the industry of Morhof and Tiraboschi[*]. The Humorists are named by Morhof as the only Italian philosophical society anterior to the Lynceans; their founder was Paolo Mancino, and the distinctive symbol which they adopted was rain dropping from a cloud, with the motto *Redit agmine dulci;*—their title is derived from the same metaphor. The object of their union appears to have been similar to that of the Lynceans, but they at no time attained to the celebrity to which Cesi's society rose from the moment of its incorporation. Cesi took the presidency for his life, and the celebrated Baptista Porta was appointed vice president at Naples. Stelluti acted as the legal representative of the society, with the title of procuratore. Of the other two original members Anastasio de Filiis was dead, and although Hecke returned to Italy in 1614, and rejoined the Academy, yet he was soon afterwards struck off the list in consequence of his lapsing into insanity. Among the academicians we find the names of Galileo, Fabio Colonna, Lucas Valerio, Guiducci, Welser, Giovanni Fabro, Terrentio, Virginio Cesarini, Ciampoli, Molitor, Cardinal Barberino, (nephew of Pope Urban VIII.) Stelliola, Salviati, &c.

The principal monument still remaining of the zeal and industry to which Cesi incited his academicians is the Phytobasanos, a compendium of the natural history of Mexico, which must be considered a surprising performance for the times in which it appeared. It was written by a Spaniard named Hernandez; and Reecho, who often has the credit of the whole work, made great additions to it. During fifty years the manuscript had been neglected, when Cesi discovered it, and employed Terrentio, Fabro, and Colonna, all Lynceans, to publish it enriched with their notes and emendations. Cesi himself published several treatises, two of which are extant; his *Tabulæ Phytosophicæ*, and a Dissertation on Bees entitled *Apiarium*, the only known copy of which last is in the library of the Vatican. His great work, *Theatrum Naturæ*, was never printed; a circumstance which tends to shew that he did not assemble the society round him for the purpose of ministering to his own vanity, but postponed the publication of his own productions to the labours of his coadjutors. This, and many other valuable works belonging to the academy existed in manuscript till lately in the Albani Library at Rome. Cesi collected, not a large, but an useful library for the use of the academy, (which was afterwards augmented on the premature death of Cesarini by the donation of his books); he filled a botanical garden with the rarer specimens of plants, and arranged a museum of natural curiosities; his palace at Rome was constantly open to the academicians; his purse and his influence were employed with equal liberality in their service.

Cesi's death, in 1632, put a sudden stop to the prosperity of the society, a consequence which may be attributed to the munificence with which he had from the first sustained it: no one could be found to fill his place in the princely manner to which the academieiaus were accustomed, and the society, after lingering some years under the nominal patronage of Urban VIII., gradually decayed, till, by the death of its principal members, and dispersion of the rest, it became entirely extinct[*]. Bianchi,

---

[*] Polyhistor Literarius, &c.—Storia della Letterat. Ital. The still existing society of Chaff, more generally known by its Italian title, Della Crusca, belongs to the same period.

[*] F. Colonnæ Phytobasanus Jano Planco Auctore. Florent, 1744.

whose sketch of the academy was almost the only one till the appearance of Odescalchi's history, made an attempt to revive it in the succeeding century, but without any permanent effect. A society under the same name has been formed since 1784, and is still flourishing in Rome. Before leaving the subject it may be mentioned, that one of the earliest notices that Bacon's works were known in Italy is to be found in a letter to Cesi, dated 1625; in which Pozzo, who had gone to Paris with Cardinal Barberino, mentions having seen them there with great admiration, and suggests that Bacon would be a fit person to be proposed as a member of their society. After Galileo's death, three of his principal followers, Viviani, Torricelli, and Aggiunti formed the plan of establishing a similar philosophical society, and though Aggiunti and Torricelli died before the scheme could be realized, Viviani pressed it forward, and, under the auspices of Ferdinand II., formed a society, which, in 1657, merged in the famous *Academia del Cimento*, or Experimental Academy. This latter held its occasional meetings at the palace of Ferdinand's brother, Leopold de' Medici: it was composed chiefly, if not entirely, of Galileo's pupils and friends. During the few years that this society lasted, one of the principal objects of which was declared to be the repetition and developement of Galileo's experiments, it kept up a correspondence with the principal philosophers in every part of Europe, but when Leopold was, in 1666, created a cardinal, it appears to have been dissolved, scarcely ten years after its institution†. This digression may be excused in favour of so interesting an establishment as the Academia Lincea, which preceded by half a century the formation of the Royal Society of London, and Académie Françoise of Paris.

These latter two are mentioned together, probably for the first time, by Salusbury. The passage is curious in an historical point of view, and worth extracting:—" In imitation of these societies, Paris and London have erected theirs of *Les Beaux Esprits*, and of the *Virtuosi*: the one by the countenance of the most eminent Cardinal Richelieu, the other by the royal encouragement of his sacred Majesty that now is. The *Beaux Esprits* have published sundry volumes of their moral and physiological conferences, with the laws and history of their fellowship; and I hope the like, in due time from our Royal Society; that so such as envie their fame and felicity, and such as suspect their ability and candor, may be silenced and disappointed in their detractions and expectations."*

CHAPTER X.

Spots on the Sun—Essay on Floating Bodies—Scheiner—Change in Saturn.

GALILEO did not indulge the curiosity of his Roman friends by exhibiting only the wonders already mentioned, which now began to lose the gloss of novelty, but disclosed a new discovery, which appeared still more extraordinary, and, to the opposite faction, more hateful than anything of which he had yet spoken. This was the discovery, which he first made in the month of March, 1611, of dark spots on the body of the sun. A curious fact, and one which well serves to illustrate Galileo's superiority in seeing things simply as they are, is, that these spots had been observed and recorded centuries before he existed, but, for want of careful observation, their true nature had been constantly misapprehended. One of the most celebrated occasions was in the year 807 of our era, in which a dark spot is mentioned as visible on the face of the sun during seven or eight days. It was then supposed to be Mercury†. Kepler, whose astronomical knowledge would not suffer him to overlook that it was impossible that Mercury could remain so long in conjunction with the sun, preferred to solve the difficulty by supposing that, in Aimoin's original account, the expression was not *octo dies* (eight days), but *octoties*—a barbarous word, which he supposed to have been written for *octies* (eight times); and that the other accounts (in which the number of days mentioned is different) copying loosely from the first, had both mistaken the word, and misquoted the time which they thought they found mentioned there. It is impossible to look on this explanation as satisfactory, but Kepler, who at that time did not dream of spots on the sun, was perfectly contented with it. In 1609, he himself observed upon the sun a black spot, which he in like manner mistook for Mercury, and unluckily the day, being cloudy, did

---

* Nelli Saggio di Storia Literaria Fiorentina, Lucca, 1759.

* Salusbury's Math. Coll. vol. ii. London, 1664.
† Aimoini Hist. Francorum. Parisiis. 1567.

not allow him to contemplate it sufficiently long to discover his error, which the slowness of its apparent motion would soon have pointed out.* He hastened to publish his supposed observation, but no sooner was Galileo's discovery of the solar spots announced, than he, with that candour which as much as his flighty disposition certainly characterized him at all times, retracted his former opinion, and owned his belief that he had been mistaken. In fact it is known from the more accurate theory which we now possess of Mercury's motions, that it did not pass over the sun's face at the time when Kepler thought he perceived it there.

Galileo's observations were in their consequences to him particularly unfortunate, as in the course of the controversy in which they engaged him, he first became personally embroiled with the powerful party, whose prevailing influence was one of the chief causes of his subsequent misfortunes. Before we enter upon that discussion, it will be proper to mention another famous treatise which Galileo produced soon after his return from Rome to Florence, in 1612. This is, his Discourse on Floating Bodies, which restored Archimedes' theory of hydrostatics, and has, of course, met with the opposition which few of Galileo's works failed to encounter. In the commencement, he thought it necessary to apologize for writing on a subject so different from that which chiefly occupied the public attention, and declared that he had been too closely occupied in calculating the periods of the revolutions of Jupiter's satellites to permit him to publish anything earlier. These periods he had succeeded in determining during the preceding year, whilst at Rome, and he now announced them to complete their circuits, the first in about 1 day, 18½ hours; the second in 3 days, 13 hours, 20 minutes; the third in 7 days, 4 hours; and the outermost in 16 days, 18 hours. All these numbers he gave merely as approximately true, and promised to continue his observations, for the purpose of correcting the results. He then adds an announcement of his recent discovery of the solar spots, "which, as they change their situation, offer a strong argument, either that the sun revolves on itself, or that, perhaps, other stars, like Venus and Mercury, revolve about it, invisible at all other times, on account of the small distance to which they are removed from him." To this he afterwards subjoined, that, by continued observation, he had satisfied himself that these solar spots were in actual contact with the surface of the sun, where they are continually appearing and disappearing; that their figures were very irregular, some being very dark, and others not so black; that one would often divide into three or four, and, at other times, two, three, or more would unite into one; besides which, that they had all a common and regular motion, with which they revolved round with the sun, which turned upon its axis in about the time of a lunar month.

Having by these prefatory observations assuaged the public thirst for astronomical novelties, he ventures to introduce the principal subject of the treatise above mentioned. The question of floating bridges had been discussed at one of the scientific parties, assembled at the house of Galileo's friend Salviati, and the general opinion of the company appearing to be that the floating or sinking of a body depended principally upon its shape, Galileo undertook to convince them of their error. If he had not preferred more direct arguments, he might merely have told them that in this instance they were opposed to their favourite Aristotle, whose words are very unequivocal on the point in dispute. "Form is not the cause why a body moves downwards rather than upwards, but it does affect the swiftness with which it moves;"* which is exactly the distinction which those who called themselves Aristotelians were unable to perceive, and to which the opinions of Aristotle himself were not always true. Galileo states the discussion to have immediately arisen from the assertion of some one in the company, that condensation is the effect of cold, and ice was mentioned as an instance. On this, Galileo observed, that ice is rather water rarefied than condensed, the proof of which is, that ice always floats upon water.† It was replied, that the reason of this phenomenon was, not the superior lightness of the ice, but its incapacity, owing to its flat shape, to penetrate and overcome the resistance of the water. Galileo denied this, and asserted that ice of any shape would float upon water, and that, if a

* Mercurius in sole visus. 1609.

* De Cœlo. lib. 4.
† For a discussion of this singular phenomenon, see Treatise on Heat, p. 12; and it is worth while to remark in passing, what an admirable instance it affords of Galileo's instantaneous abandonment of a theory so soon as it became inconsistent with experiment.

flat piece of ice were forcibly taken to the bottom, it would of itself rise again to the surface. Upon this assertion it appears that the conversation became so clamorous, that Galileo thought it pertinent to commence his Essay with the following observation on the advantage of delivering 'scientific opinions in writing, " because in conversational arguments, either one or other party, or perhaps both, are apt to get overwarm, and to speak overloud, and either do not suffer each other to be heard, or else, transported with the obstinacy of not yielding, wander far away from the original proposition, and confound both themselves and their auditors with the novelty and variety of their assertions." After this gentle rebuke he proceeds with his argument, in which he takes occasion to state the famous hydrostatical paradox, of which the earliest notice is to be found in Stevin's works, a contemporary Flemish engineer, and refers it to a principle on which we shall enlarge in another chapter. He then explains the true theory of buoyancy, and refutes the false reasoning on which the contrary opinions were founded, with a variety of experiments.

The whole value and interest of experimental processes generally depends on a variety of minute circumstances, the detail of which would be particularly unsuited to a sketch like the present one. For those who are desirous of becoming more familiar with Galileo's mode of conducting an argument, it is fortunate that such a series of experiments exists as that contained in this essay; experiments which, from their simplicity, admit of being for the most part concisely enumerated, and at the same time possess so much intrinsic beauty and characteristic power of forcing conviction. They also present an admirable specimen of the talent for which Galileo was so deservedly famous, of inventing ingenious arguments in favour of his adversaries' absurd opinions before he condescended to crush them, shewing that nothing but his love of truth stood in the way of his being a more subtle sophist than any amongst them. In addition to these reasons for giving these experiments somewhat in detail, is the fact that all explanation of one of the principal phenomena to which they allude is omitted in many more modern treatises on Hydrostatics; and in some it is referred precisely to the false doctrines here confuted.

The marrow of the dispute is included in Galileo's assertion, that "The diversity of figure given to any solid cannot be in any way the cause of its absolutely sinking or floating; so that if a solid, when formed for example into a spherical figure, sinks or floats in the water, the same body will sink or float in the same water, when put into any other form. The breadth of the figure may indeed retard its velocity, as well of ascent as descent, and more and more according as the said figure is reduced to a greater breadth and thinness; but that it may be reduced to such a form as absolutely to put an end to its motion in the same fluid, I hold to be impossible. In this I have met with great contradictors who, producing some experiments, and in particular a thin board of ebony, and a ball of the same wood, and shewing that the ball in water sinks to the bottom*, and that the board if put lightly on the surface floats, have held and confirmed themselves in their opinion with the authority of Aristotle, that the cause of that rest is the breadth of the figure, unable by its small weight to pierce and penetrate the resistance of the water's thickness, which is readily overcome by the other spherical figure."—For the purpose of these experiments, Galileo recommends a substance such as wax, which may be easily moulded into any shape, and with which, by the addition of a few filings of lead, a substance may be readily made of any required specific gravity. He then declares that if a ball of wax of the size of an orange, or bigger, be made in this manner heavy enough to sink to the bottom, but so lightly that if we take from it only one grain of lead it returns to the top; and if the same wax be afterwards moulded into a broad and thin cake, or into any other figure, regular or irregular, the addition of the same grain of lead will always make it sink, and it will again rise when we remove the lead from it.—" But methinks I hear some of the adversaries raise a doubt upon my produced experiment: and, first, they offer to my consideration that the figure, as a figure simply, and disjunct from the matter, works no effect, but requires to be conjoined with the matter; and, moreover, not with every matter, but with those only wherewith it may be able to execute the desired operation. Just as we see by experience

---

* Ebony is one of the few woods heavier than water. See Treatise on Hydrostatics.

that an acute and sharp angle is more apt to cut than an obtuse; yet always provided that both one and the other are joined with a matter fit to cut, as for instance, steel. Therefore a knife with a fine and sharp edge cuts bread or wood with much ease, which it will not do if the edge be blunt and thick; but if, instead of steel, any one will take wax and mould it into a knife, undoubtedly he will never learn the effects of sharp and blunt edges, because neither of them will cut; the wax being unable, by reason of its flexibility, to overcome the hardness of the wood and bread. And therefore, applying the like discourse to our argument, they say that the difference of figure will shew different effects with regard to floating and sinking, but not conjoined with any kind of matter, but only with those matters which by their weight are able to overcome the viscosity of the water (like the ebony which they have selected); and he that will select cork or other light wood to form solids of different figures, would in vain seek to find out what operation figure has in sinking or floating, because all would swim, and that not through any property of this or that figure, but through the debility of the matter."

." When I begin to examine one by one all the particulars here produced, I allow not only that figures, simply as such, do not operate in natural things, but also that they are never separated from the corporeal substance, nor have I ever alleged them to be stript of sensible matter: and also I freely admit, that in our endeavours to examine the diversity of accidents which depend upon the variety of figures, it is necessary to apply them to matters which obstruct not the various operations of those various figures. I admit and grant that I should do very ill if I were to try the influence of a sharp edge with a knife of wax, applying it to cut an oak, because no sharpness in wax is able to cut that very hard wood. But yet, such an experiment of this knife would not be beside the purpose to cut curded milk, or other very yielding matter; nay, in such matters, the wax is more convenient than steel for finding the difference depending on the acuteness of the angles, because milk is cut indifferently with a razor, or a blunt knife. We must therefore have regard not only to the hardness, solidity, or weight of the bodies which, under different figures, are to divide some matters asunder; but also, on the other hand, to the resistance of the matter to be penetrated. And, since I have chosen a matter which does penetrate the resistance of the water, and in all figures descends to the bottom, my antagonists can charge me with no defect; nor (to revert to their illustration) have I attempted to test the efficacy of acuteness by cutting with matters unable to cut. I subjoin withal, that all caution, distinction, and election of matter would be superfluous and unnecessary, if the body to be cut should not at all resist the cutting: if the knife were to be used in cutting a mist, or smoke, one of paper would serve the purpose as well as one of Damascus steel; and I assert that this is the case with water, and that there is not any solid of such lightness or of such a figure, that being put on the water it will not divide and penetrate its thickness; and if you will examine more carefully your thin boards of wood, you will see that they have part of their thickness under water; and, moreover, you will see that the shavings of ebony, stone, or metal, when they float, have not only thus broken the continuity of the water, but are with all their thickness under the surface of it; and that more and more, according as the floating substance is heavier, so that a thin floating plate of lead will be lower than the surface of the surrounding water by at least twelve times the thickness of the plate, and gold will dive below the level of the water almost twenty times the thickness of the plate, as I shall shew presently."

In order to illustrate more clearly the non-resistance of water to penetration, Galileo then directs a cone to be made of wood or wax, and asserts that when it floats, either with its base or point in the water, the solid content of the part immersed will be the same, although the point is, by its shape, better adapted to overcome the resistance of the water to division, if that were the cause of the buoyancy. Or the experiment may be varied by tempering the wax with filings of lead, till it sinks in the water, when it will be found that in any figure the same cork must be added to it to raise it to the surface.—
"This silences not my antagonists; but they say that all the discourse hitherto made by me imports little to them, and that it serves their turn, that they have demonstrated in one instance, and in such manner and figure as pleases them best, namely, in a board and a ball of ebony;

that one, when put into the water, sinks to the bottom, and that the other stays to swim at the top; and the matter being the same, and the two bodies differing in nothing but in figure, they affirm that with all perspicuity they have demonstrated and sensibly manifested what they undertook. Nevertheless I believe, and think I can prove that this very experiment proves nothing against my theory. And first it is false that the ball sinks, and the board not; for the board will sink too, if you do to both the figures as the words of our question require; that is, if you put them both *in* the water; for to be in the water implies to be placed in the water, and by Aristotle's own definition of place, to be placed imports to be environed by the surface of the ambient body; but when my antagonists shew the floating board of ebony, they put it not into the water, but upon the water; where, being detained by a certain impediment (of which more anon) it is surrounded, partly with water, partly with air, which is contrary to our agreement, for that was that the bodies should be in the water, and not part in the water, part in the air. I will not omit another reason, founded also upon experience, and, if I deceive not myself, conclusive against the notion that figure, and the resistance of the water to penetration have anything to do with the buoyancy of bodies. Choose a piece of wood or other matter, as for instance walnutwood, of which a ball rises from the bottom of the water to the surface more slowly than a ball of ebony of the same size sinks, so that clearly the ball of ebony divides the water more readily in sinking than does the walnut in rising. Then take a board of walnut-tree equal to and like the floating ebony one of my antagonists; and if it be true that this latter floats by reason of the figure being unable to penetrate the water, the other of walnut-tree, without all question, if thrust to the bottom ought to stay there, as having the same impeding figure, and being less apt to overcome the said resistance of the water. But if we find by experience that not only the thin board, but every other figure of the same walnut-tree will return to float, as unquestionably we shall, then I must desire my opponents to forbear to attribute the floating of the ebony to the figure of the board, since the resistance of the water is the same in rising as in sinking, and the force of ascension of the walnut-tree is less than the ebony's force for going to the bottom."

"Now, let us return to the thin plate of gold or silver, or the thin board of ebony, and let us lay it lightly upon the water, so that it may stay there without sinking, and carefully observe the effect. It will appear clearly that the plates are a considerable matter lower than the surface of the water which rises up, and makes a kind of rampart round them on every side, in the manner shewn in the annexed figure, in which B D L F repre-

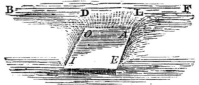

sents the surface of the water, and A E I O the surface of the plate. But if it have already penetrated and overcome the continuity of the water, and is of its own nature heavier than the water, why does it not continue to sink, but stop and suspend itself in that little dimple that its weight has made in the water? My answer is, because in sinking till its surface is below the water which rises up in a hauk round it, it draws after and carries along with it the air above it, so that that which in this case descends and is placed in the water, is not only the board of ebony or plate of iron, but a compound of ebony and air, from which composition results a solid no longer specifically heavier than the water, as was the ebony or gold alone. But, Gentlemen, we want the same matter; you are to alter nothing but the shape, and therefore have the goodness to remove this air, which may be done simply by washing the upper surface of the board, for the water having once got between the board and air will run together, and the ebony will go to the bottom; and if it does not, you have won the day. But methinks I hear some of my antagonists cunningly opposing this, and telling me that they will not on any account allow their board to be wetted, because the weight of the water so added, by making it heavier than it was before, draws it to the bottom, and that the addition of new weight is contrary to our agreement, which was that the matter should be the same."

"To this I answer first, that nobody can suppose bodies to be put into the water without their being wet, nor do I wish to do more to the board than you may do to the ball. Moreover, it is not true that the board sinks on account of the weight of the water added in the washing; for I will put ten or twenty drops on the floating board, and so long as they stand separate it shall not sink; but if the board be taken out, and all that water wiped off, and the whole surface bathed with one single drop, and put it again upon the water, there is no question but it will sink, the other water running to cover it, being no longer hindered by the air. In the next place it is altogether false that water can in any way increase the weight of bodies immersed in it, for water has no weight in water, since it does not sink. Now, just as he who should say that brass by its own nature sinks, but that when formed into the shape of a kettle, it acquires from that figure a virtue of lying in the water without sinking, would say what is false, because that is not purely brass which then is put into the water, but a compound of brass and air; so is it neither more nor less false, that a thin plate of brass or ebony swims by virtue of its dilated and broad figure. Also I cannot omit to tell my opponents, that this conceit of refusing to bathe the surface of the board, might beget an opinion in a third person of a poverty of arguments on their side, especially as the conversation began about flakes of ice, in which it would be simple to require that the surfaces should be kept dry; not to mention that such pieces of ice, whether wet or dry, always float, and as my antagonists say, because of their shape."

"Some may wonder that I affirm this power to be in the air of keeping the plate of brass or silver above water, as if in a certain sense I would attribute to the air a kind of magnetic virtue for sustaining heavy bodies with which it is in contact. To satisfy all these doubts, I have contrived the following experiment to demonstrate how truly the air does support these solids; for I have found, when one of these bodies which floats when placed lightly on the water, is thoroughly bathed and sunk to the bottom, that by carrying down to it a little air without otherwise touching it in the least, I am able to raise and carry it back to the top, where it floats as before. To this effect I take a ball of wax, and with a little lead make it just heavy enough to sink very slowly to the bottom, taking care that its surface be quite smooth and even. This, if put gently into the water, submerges almost entirely, there remaining visible only a little of the very top, which, so long as it is joined to the air, keeps the ball afloat; but if we take away the contact of the air by wetting this top, the ball sinks to the bottom, and remains there. Now to make it return to the surface by virtue of the air which before sustained it, thrust into the water a glass, with the mouth downwards, which will carry with it the air it contains; and move this down towards the ball, until you see by the transparency of the glass that the air has reached the top of it; then gently draw the glass upwards, and you will see the ball rise, and afterwards stay on the top of the water, if you carefully part the glass and water without too much disturbing it*. There is therefore a certain affinity between the air and other bodies, which holds them united, so that they separate not without a kind of violence, just as between water and other bodies; for in drawing them wholly out of the water, we see the water follow them, and rise sensibly above the level before it quits them." Having established this principle by this exceedingly ingenious and convincing experiment, Galileo proceeds to shew from it what must be the dimensions of a plate of any substance which will float as the wax does, assuming in each case that we know the greatest height at which the rampart of water will stand round it. In like manner he shows that a pyramidal or conical figure may be made of any substance, such that by help of the air, it shall rest upon the water without wetting more than its base; and that we may so form a cone of any substance that it shall float if placed gently on the surface, with its point downwards, whereas no care or pains will enable it to float with its base downwards, owing to the different proportions of air which in the two positions remain connected with it. With this parting blow at his antagonist's theory we close our extracts from this admirable essay.

The first elements of the theory of running waters were reserved for Castelli, an intimate friend and pupil of Galileo. On the present occasion, Castelli appeared as the ostensible author of a de-

---

* In making this very beautiful experiment, it is best to keep the glass a few seconds in the water, to give time for the surface of the ball to dry. It will also succeed with a light needle, if carefully conducted.

fence against the attacks made by Vincenzio di Grazia and by Lodovico delle Columbe (the author of the crystalline composition of the moon) on the obnoxious theory. After destroying all the objections which they produced, the writer tauntingly bids them remember, that he was merely Galileo's pupil, and consider how much more effectually Galileo himself would have confuted them, had he thought it worth while. It was not known till several years after his death, that this Essay was in fact written by Galileo himself.*

These compositions merely occupied the leisure time which he could withhold from the controversy on the solar spots to which we have already alluded. A German Jesuit named Christopher Scheiner, who was professor of mathematics at Ingolstadt, in imitation of Galileo had commenced a series of observations on them, but adopted the theory which, as we have seen, Galileo had examined and rejected, that these spots are planets circulating at some distance from the body of the sun. The same opinion had been taken up by a French astronomer, who in honour of the reigning family called them Borbonian stars. Scheiner promulgated his notions in three letters, addressed to their common friend Welser, under the quaint signature of "*Apelles latens post tabulam.*" Galileo replied to Scheiner's letters by three others, also addressed to Welser, and although the dispute was carried on amid mutual professions of respect and esteem, it laid the foundation of the total estrangement which afterwards took place between the two authors. Galileo's part of this controversy was published at Rome by the Lyncean Academy in 1613. To the last of his letters, written in December, 1612, is annexed a table of the expected positions of Jupiter's satellites during the months of March and April of the following year, which, imperfect as it necessarily was, cannot be looked upon without the greatest interest.

In the same letter it is mentioned that Saturn presented a novel appearance, which, for an instant, almost induced Galileo to mistrust the accuracy of his earlier observations. The lateral appendages of this planet had disappeared, and the accompanying extract will show the uneasiness which Galileo could not conceal at the sight of this phenomenon, although it is admirable to see the contempt with which, even in that trying moment, he expresses his consciousness that his adversaries were unworthy of the triumph they appeared on the point of celebrating.—" Looking on Saturn within these few days, I found it solitary, without the assistance of its accustomed stars, and in short, perfectly round and defined like Jupiter, and such it still remains. Now what can be said of so strange a metamorphosis? are perhaps the two smaller stars consumed, like the spots on the sun? have they suddenly vanished and fled? or has Saturn devoured his own children? or was the appearance indeed fraud and illusion, with which the glasses have for so long a time mocked me, and so many others who have often observed with me. Now perhaps the time is come to revive the withering hopes of those, who, guided by more profound contemplations, have fathomed all the fallacies of the new observations and recognised their impossibility! I cannot resolve what to say in a chance so strange, so new, and so unexpected; the shortness of the time, the unexampled occurrence, the weakness of my intellect, and the terror of being mistaken, have greatly confounded me." These first expressions of alarm are not to be wondered at; however, he soon recovered courage, and ventured to foretel the periods at which the lateral stars would again show themselves, protesting at the same time, that he was in no respect to be understood as classing this prediction among the results which depend on certain principles and sound conclusions, but merely on some conjectures which appeared to him probable. From one of the Dialogues on the System, we learn that this conjecture was, that Saturn might revolve upon his axis, but the period which he assumed is very different from the true one, as might be expected from its being intended to account for a phenomenon of which Galileo had not rightly apprehended the character.

He closed this letter with renewed professions of courtesy and friendship towards Apelles, enjoining Welser not to communicate it without adding his excuses, if he should be thought to dissent too violently from his antagonist's ideas, declaring that his only object was the discovery of truth, and that he had freely exposed his own opinion, which he was still ready to change, so soon as his errors should be made manifest to him;

* Nelli. Saggio di Stor. Liter. Fiorent.

and that he would consider himself under special obligation to any one who would be kind enough to discover and correct them. These letters were written from the villa of his friend Salviati at Selve near Florence, where he passed great part of his time, particularly during his frequent indispositions, conceiving that the air of Florence was prejudicial to him. Cesi was very anxious for their appearance, since they were (in his own words) so hard a morsel for the teeth of the Peripatetics, and he exhorted Galileo, in the name of the society, "to continue to give them, and the nameless Jesuit, something to gnaw."

CHAPTER XI.

Letter to Christina, Arch-Duchess of Tuscany — Caccini — Galileo revisits Rome — Inchofer — Problem of Longitudes.

THE uncompromising boldness with which Galileo published and supported his opinions, with little regard to the power and authority of those who advocated the contrary doctrines, had raised against him a host of enemies, who each had objections to him peculiar to themselves, but who now began to perceive the policy of uniting their strength in the common cause, to crush if possible so dangerous an innovator. All the professors of the old opinions, who suddenly found the knowledge on which their reputation was founded struck from under them, and who could not reconcile themselves to their new situation of learners, were united against him; and to this powerful cabal was now added the still greater influence of the jesuits and pseudo-theological party, who fancied they saw in the spirit of Galileo's writings the same inquisitive temper which they had already found so inconvenient in Luther and his adherents. The alarm became greater every day, inasmuch as Galileo had succeeded in training round him a numerous band of followers who all appeared imbued with the same dangerous spirit of innovation, and his favourite scholars were successful candidates for professorships in many of the most celebrated universities of Italy.

At the close of 1613, Galileo addressed a letter to his pupil, the Abbé Castelli, in which he endeavoured to shew that there is as much difficulty in reconciling the Ptolemaic as the Copernican system of the world with the astronomical expressions contained in the Scriptures, and asserted, that the object of the Scriptures not being to teach astronomy, such expressions are there used as would be intelligible and conformable to the vulgar belief, without regard to the true structure of the universe; which argument he afterwards amplified in a letter addressed to Christina, Grand Duchess of Tuscany, the mother of his patron Cosmo. He discourses on this subject with the moderation and good sense which so peculiarly characterized him. "I am," says he, "inclined to believe, that the intention of the sacred Scriptures is to give to mankind the information necessary for their salvation, and which, surpassing all human knowledge, can by no other means be accredited than by the mouth of the Holy Spirit. But I do not hold it necessary to believe, that the same God who has endowed us with senses, with speech, and intellect, intended that we should neglect the use of these, and seek by other means for knowledge which they are sufficient to procure us; especially in a science like astronomy, of which so little notice is taken in the Scriptures, that none of the planets, except the sun and moon, and, once or twice only, Venus under the name of Lucifer, are so much as named there. This therefore being granted, methinks that in the discussion of natural problems we ought not to begin at the authority of texts of Scripture, but at sensible experiments and necessary demonstrations: for, from the divine word, the sacred Scripture and nature did both alike proceed, and I conceive that, concerning natural effects, that which either sensible experience sets before our eyes, or necessary demonstrations do prove unto us, ought not upon any account to be called into question, much less condemned, upon the testimony of Scriptural texts, which may under their words couch senses seemingly contrary thereto.

"Again, to command the very professors of astronomy that they of themselves see to the confuting of their own observations and demonstrations, is to enjoin a thing beyond all possibility of doing; for it is not only to command them not to see that which they do see, and not to understand that which they do understand, but it is to order them to seek for and to find the contrary of that which they happen to meet with. I would entreat these wise and prudent fathers, that they would with all diligence cousider the difference that is between opininative and demonstrative doctrines: to the end that well weighing in their minds with what force necessary inferences urge us, they might the better assure themselves that it is not in the power of the professors of demonstrative sciences to change their opinions at pleasure, and adopt first one side and then another; and that there is a great difference between commanding a mathematician or a philosopher, and the disposing of a lawyer or a merchant; and that the demonstrated conclusions touching the things of nature and of the heavens cannot be changed with the same facility as the opinions are touching what is lawful or not in a contract, bargain, or bill of exchange. Therefore, first let these men apply themselves to examine the arguments of Copernicus and others, and leave the condemning of them as erroneous and heretical to whom it belongeth; yet let them not hope to find such rash and precipitous determinations in the wary and holy fathers, or in the absolute wisdom of him who cannot err, as those into which they suffer themselves to be hurried by some particular affection or interest of their own. In these and such other positions, which are not directly articles of faith, certainly no man doubts but His Holiness hath always an absolute power of admitting or condemning them, but it is not in the power of any creature to make them to be true or false, otherwise than of their own nature, and in fact they are." We have been more particular in extracting these passages, because it has been advanced by a writer of high reputation, that the treatment which Galileo subsequently experienced was solely in consequence of his persisting in the endeavour to prove that the Scriptures were reconcileable with the Copernican theory*, whereas we see here distinctly that, for the reasons we have briefly stated, he regarded this as a matter altogether indifferent and beside the question.

'Galileo had not entered upon this discussion till driven to it by a most indecent attack, made on him from the pulpit, by a Dominican friar named Caccini, who thought it not unbecoming his habit or religion to play upon the words of a Scriptural text for the purpose of attacking Galileo and his partisans with more personality*. Galileo complained formally of Caccini's conduct to Luigi Maraffi the general of the Dominicans, who apologised amply to him, adding that he himself was to be pitied for finding himself implicated in all the brutal conduct of thirty or forty thousand monks.

In the mean time, the inquisitors at Rome had taken the alarm, and were already, in 1615, busily employed in colleeting evidence against Galileo. Lorini, a brother Dominican of Caccini, had given them notice of the letter to Castelli of which we have spoken, and the utmost address was employed to get the original into their hands, which attempt however was frustrated, as Castelli had returned it to the writer. Caccini was sent for to Rome, settled there with the title of Master of the Convent of St. Mary of Minerva, and employed to put the depositions against Galileo into order. Galileo was not at this time fully aware of the machinations against him, but suspecting something of their nature, he solicited and obtained permission from Cosmo, towards the end of 1615, to make a journey to Rome, for the purpose of more directly confronting his enemies in that city. There was a rumour at the time that this visit was not voluntary, but that Galileo had been cited to appear at Rome. A contemporary declares that he heard this from Galileo himself: at any rate, in a letter which Galileo shortly afterwards wrote to Picchena, the Grand Duke's secretary, he expresses himself well satisfied with the results of this step, whether forced or not, and Querenghi thus describes to the Cardinal d'Este the public effect of his appearance: "Your Eminence would be delighted with Galileo if you heard him holding forth, as he often does, in the midst of fifteen or twenty, all violently attacking him, sometimes in one house, sometimes in another. But he is armed after such fashion that he laughs all of them to scorn—and even if the novelty of his opinions prevents entire persuasion, at least he convicts of emptiness most of the arguments with which his adversaries endeavour to overwhelm him. He was particularly admi-

* Ce philosophe (Galilée) ne fut point persecuté comme bon astronome, mais comme mauvais théologien. C'est son entêtement à vouloir concilier la Bible avec Copernic qui lui donna des juges. Mais vingt auteurs, surtout parmi les protestans, ont écrit que Galilée fut persecuté et imprisonné pour avoir soutenu que la terre tourne autour du soleil, que ce système a été condamné par l'inquisition comme faux, erroné et contraire à la Bible, &c.—Bergier, Encyclopédie Méthodique, Paris, 1790, Art. SCIENCES HUMAINES.

* Viri Galilæi, quid statis adspicientes in cœlum. Acts I. II.

rable on Monday last, in the house of Signor Frederico Ghisilieri; and what especially pleased me was, that before replying to the contrary arguments, he amplified and enforced them with new grounds of great plausibility, so as to leave his adversaries in a more ridiculous plight when he afterwards overturned them all."

Among the malicious stories which were put into circulation, it had been said, that the Grand Duke had withdrawn his favour, which emboldened many, who would not otherwise have ventured on such open opposition, to declare against Galileo. His appearance at Rome, where he was lodged in the palace of Cosmo's ambassador, and whence he kept up a close correspondence with the Grand Duke's family, put an immediate stop to rumours of this kind. In little more than a month he was apparently triumphant, so far as regarded himself; but the question now began to be agitated whether the whole system of Copernicus ought not to be condemned as impious and heretical. Galileo again writes to Picchena, "so far as concerns the clearing of my own character, I might return home immediately; but although this new question regards me no more than all those who for the last eighty years have supported these opinions both in public and private, yet, as perhaps I may be of some assistance in that part of the discussion which depends on the knowledge of truths ascertained by means of the sciences which I profess, I, as a zealous and Catholic Christian, neither can nor ought to withhold that assistance which my knowledge affords; and this business keeps me sufficiently employed." De Lambre, whose readiness to depreciate Galileo's merit we have already noticed and lamented, sneeringly and ungratefully remarks on this part of his life; that " it was scarcely worth while to compromise his tranquillity and reputation, in order to become the champion of a truth which could not fail every day to acquire new partisans by the natural effect of the progress of enlightened opinions." We need not stop to consider what the natural effects might have been if none had at any time been found who thought their tranquillity worthily offered up in such a cause.

It has been hinted by several, and is indeed probable, that Galileo's stay at Rome rather injured the cause (so far as provoking the inquisitorial censures could injure it) which it was his earnest desire to serve, for we cannot often enough repeat the assertion, that it was not the doctrine itself, so much as the free, unyielding manner in which it was supported, which was originally obnoxious. Copernicus had been allowed to dedicate his great work to Pope Paul III., and from the time of its first appearance under that sanction in 1543, to the year 1616, of which we are now writing, this theory was left in the hands of mathematicians and philosophers, who alternately attacked and defended it without receiving either support or molestation from ecclesiastical decrees. But this was henceforward no longer the case, and a higher degree of importance was given to the controversy from the religions heresies which were asserted to be involved in the new opinions. We have already given specimens of the so called philosophical arguments brought against Copernicus; and the reader may be curious to know the form of the theological ones. Those which we select are taken from a work, which indeed did not come forth till the time of Galileo's third visit to Rome, but it is relative to the matter now before us, as it professed to be, and its author's party affected to consider it, a complete refutation of the letters to Castelli and the Archduchess Christina*.

It was the work of a Jesuit, Melchior Inchoffer, and it was greatly extolled by his companions, " as differing so entirely from the pruriency of the Pythagorean writings." He quotes with approbation an author who, first referring to the first verse of Genesis for an argument that the earth was not created till after the heavens, observes that the whole question is thus reduced to the examination of this purely geometrical difficulty—In the formation of a sphere, does the centre or circumference first come into existence? If the latter (which we presume Melchior's friend found good reason for deciding upon), the cousequence is inevitable. The earth is in the centre of the universe.

It may not be unprofitable to contrast the extracts which we have given from Galileo's letters on the same subject with the following passage, which appears one of the most subtle and argumen-

* Tractatus Syllepticus. Romæ, 1633. ' The title-page of this remarkable production is decorated with an emblematical figure, representing the earth included in a triangle; and in the three corners, grasping the globe with their fore feet, are placed three bees, the arms of Pope Urban VIII. who condemned Galileo and his writings. The motto is " His fixa quiescit," " Fixed by these it is at rest."

tative which is to be found in Melchior's book. He *professes* to be enumerating and refuting the principal arguments which the Copernicans adduced for the motion of the earth. "Fifth argument. Hell is in the centre of the earth, and in it is a fire tormenting the damned; therefore it is absolutely necessary that the earth is moveable. The antecedent is plain." (Inchoffer then quotes a number of texts of Scripture on which, according to him, the Copernicans relied in proof of this part of the argument.) "The consequent is proved: because fire is the cause of motion, for which reason Pythagoras, who, as Aristotle reports, puts the place of punishment in the centre, perceived that the earth is animate and endowed with action. I answer, even allowing that hell is in the centre of the earth, and a fire in it, I deny the consequence: and for proof I say, if the argument is worth any thing, it proves also that lime-kilns, ovens, and fire-grates are animate and spontaneously moveable. I say, *even allowing* that hell is in the centre of the earth: for Gregory, book 4, dial. chap. 42, says, that he dare not decide rashly on this matter, although he thinks more probable the opinion of those who say that it is under the earth. St. Thomas, in Opusc. 10, art. 31, says: Where hell is, whether in the centre of the earth or at the surface, does not in my opinion, relate to any article of faith; and it is superfluous to be solicitous about such things, either in asserting or denying them. And Opusc. 11, art. 24, he says, that it seems to him that nothing should be rashly asserted on this matter, particularly as Augustin thinks that nobody knows where it is; but I do not, says he, think that it is in the centre of the earth. I should be loth, however, that it should be hence inferred by *some people* that hell is in the earth, that we are ignorant where hell is, and therefore that the situation of the earth is also unknown, and, in conclusion, that it cannot therefore be the centre of the universe. The argument shall be retorted in another fashion: for if the place of the earth is unknown, it cannot be said to be in a great circle, so as to be moved round the sun. Finally I say that in fact it is known where the earth is."

It is not impossible that some persons adopted the Copernican theory, from an affectation of singularity and freethinking, without being able to give very sound reasons for their change of opinion, of whom we have an instance in Origanus, the astrological instructor of Wallenstein's famous attendant Seni, who edited his work. His arguments in favour of the earth's motion are quite on a level with those advanced on the opposite side in favour of its immobility; but we have not found any traces whatever of such absurdities as these having been urged by any of the leaders of that party, and it is far more probable that they are the creatures of Melchior's own imagination. At any rate it is worth remarking how completely he disregards the real physical arguments, which he ought, in justice to his cause, to have attempted to controvert. His book was aimed at Galileo and his adherents, and it is scarcely possible that he could seriously persuade himself that he was stating and overturning arguments similar to those by which Galileo had made so many converts to the opinions of Copernicus. Whatever may be our judgment of his candour, we may at least feel assured that if this had indeed been a fair specimen of Galileo's philosophy, he might to the end of his life have taught that the earth moved round the sun, or if his fancy led him to a different hypothesis, he might like the Abbé Baliani have sent the earth spinning round the stationary moon, and like him have remained unmolested by pontifical censures. It is true that Baliani owned his opinion to be much shaken, on observing it to be opposed to the decree of those in whose hands was placed the power of judging articles of faith. But Galileo's uncompromising spirit of analytical investigation, and the sober but invincible force of reasoning with which he beat down every sophism opposed to him, the instruments with which he worked, were more odious than the work itself, and the condemnation which he had vainly hoped to avert was probably on his very account accelerated.

Galileo, according to his own story, had in March 1616 a most gracious audience of the pope, Paul V., which lasted for nearly an hour, at the end of which his holiness assured him, that the Congregation were no longer in a humour to listen lightly to calumnies against him, and that so long as he occupied the papal chair, Galileo might think himself out of all danger. But nevertheless he was not allowed to return home, without receiving formal notice not to teach the opinions of Co-

E

pernicus, that the sun is in the centre of the system, and that the earth moves about it, from that time forward, in any manner. That these were the literal orders given to Galileo will be presently proved from the recital of them in the famous decree against him, seventeen years later. For the present, his letters which we have mentioned, as well as one of a similar tendency by Foscarini, a Carmelite friar—a commentary on the book of Joshua by a Spaniard named Diego Zuniga—Kepler's Epitome of the Copernican Theory—and Copernicus's own work, were inserted in the list of forbidden books, nor was it till four years afterwards, in 1620, that, on reconsideration, Copernicus was allowed to be read with certain omissions and alterations then decided upon.

Galileo quitted Rome scarcely able to conceal his contempt and indignation. Two years afterwards this spirit had but little subsided, for in forwarding to the Archduke Leopold his Theory of the Tides, he accompanied it with the following remarks:—"This theory occurred to me when in Rome, whilst the theologians were debating on the prohibition of Copernicus's book, and of the opinion maintained in it of the motion of the earth, which I at that time believed; until it pleased those gentlemen to suspend the book, and declare the opinion false and repugnant to the Holy Scriptures. Now, as I know how well it becomes me to obey and believe the decisions of my superiors, which proceed out of more profound knowledge than the weakness of my intellect can attain to, this theory which I send you, which is founded on the motion of the earth, I now look upon as a fiction and a dream, and beg your highness to receive it as such. But, as poets often learn to prize the creations of their fancy, so, in hke manner, do I set some value on this absurdity of mine. It is true that when I sketched this little work, I did hope that Copernicus would not, after 80 years, be convicted of error, and I had intended to develope and amplify it farther; but a voice from heaven suddenly awakened me, and at once annihilated all my confused and entangled fancies."

It might have been predicted, from the tone of this letter alone, that it would not be long before Galileo would again bring himself under the censuring notice of the astronomical hierarchy, and indeed he had, so early as 1610, collected some of the materials for the work which caused the final explosion, and on which he now employed himself with as little intermission as the weak state of his health permitted.

He had been before this time engaged in a correspondence with the court of Spain, on the method of observing longitudes at sea, for the solution of which important problem Philip III. had offered a considerable reward, an example which has since been followed in our own and other countries. Galileo had no sooner discovered Jupiter's satellites, than he recognized the use which might be made of them for that purpose, and devoted himself with peculiar assiduity to acquiring as perfect a knowledge as possible of their revolutions. The reader will easily understand how they were to be used, if their motion could be so well ascertained as to enable Galileo at Florence to predict the exact times at which any remarkable configurations would occur, as, for instance, the times at which any one of them would be eclipsed by Jupiter. A mariner who in the middle of the Atlantic should observe the same eclipse, and compare the time of night at which he made the observation (which he might know by setting his watch by the sun on the preceding day) with the time mentioned in the predictions, would, from the difference between the two, learn the difference between the hour at Florence and the hour at the place where the ship at that time happened to be. As the earth turns uniformly round through $360°$ of longitude in 24 hours, that is, through $15°$ in each hour, the hours, minutes, and seconds of time which express this difference must be multiplied by 15, and the respective products will give the degrees, minutes, and seconds of longitude, by which the ship was then distant from Florence. This statement is merely intended to give those who are unacquainted with astronomy, a general idea of the manner in which it was proposed to use these satellites. Our moon had already been occasionally employed in the same way, but the comparative frequency of the eclipses of Jupiter's moons, and the suddenness with which they disappear, gives a decided advantage to the new method. Both methods were embarrassed by the difficulty of observing the eclipses at sea. In addition to this, it was requisite, in both methods, that the sailors should be provided with accurate means of knowing the hour, wherever they might chance to be, which was far from being the case, for although (in order not to interrupt the explanation) we have above spoken of their *watches*, yet the watches and clocks of that day were not such as could be relied on sufficiently, during the interval which must necessarily occur between the two observations. This consideration led Galileo to reflect on the use which might be made of his pendulum for this purpose; and, with respect to the other difficulty, he contrived a peculiar kind of telescope, with which he flattered himself, somewhat prematurely, that it would be as easy to observe on ship-board as on shore.

During his stay at Rome, in 1615, and the following year, he disclosed some of these ideas to the Conte di Lemos, the viceroy of Naples, who had been president of the council of the Spanish Indies, and was fully aware of the importance of the matter. Galileo was in consequence invited to communicate directly with the Duke of Lerma, the Spanish minister, and instructions were accordingly sent by Cosmo, to the Conte Orso d'Elci, his ambassador at Madrid, to conduct the business there. Galileo entered warmly into the design, of which he had no other means of verifying the practicability; for as he says in one of his letters to Spain—" Your excellency may well believe that if this were an undertaking which I could conclude by myself, I would never have gone about begging favours from others; but in my study there are neither seas, nor Indies, nor islands, nor ports, nor shoals, nor ships, for which reason I am compelled to share the enterprise with great personages, and to fatigue myself to procure the acceptance of that, which ought with eagerness to be asked of me; but I console myself with the reflection that I am not singular in this, but that it commonly happens, with the exception of a little reputation, and that too often obscured and blackened by envy, that the least part of the advantage falls to the share of the inventors of things, which afterwards bring great gain, honours, and riches to others; so that I will never cease on my part to do every thing in my power, and I am ready to leave here all my comforts, my friends, and family, and to cross over into Spain, to stay as long as I may be wanted in Seville, or Lisbon, or wherever it may be convenient, to implant the knowledge of this method, provided that due assistance and diligence be not wanting on the part of those who are to receive it, and who should solicit and foster it." But he could not, with all his enthusiasm, rouse the attention of the Spanish court. The negotiation languished, and although occasionally renewed during the next ten or twelve years, was never brought to a satisfactory issue. Some explanation of this otherwise unaccountable apathy of the Spanish court, with regard to the solution of a problem which they had certainly much at heart, is given in Nelli's life of Galileo; where it is asserted, on the authority of the Florentine records, that Cosmo required privately from Spain, (in return for the permission granted for Galileo to leave Florence, in pursuance of this design,) the privilege of sending every year from Leghorn two merchantmen, duty free, to the Spanish Indies.

CHAPTER XII.

Controversy on Comets—Saggiatore—Galileo's reception by Urban VIII—His family.

THE year 1618 was remarkable for the appearance of three comets, on which almost every astronomer in Europe found something to say and write. Galileo published some of his opinions with respect to them, through the medium of Mario Guiducci. This astronomer delivered a lecture before the Florentine academy, the heads of which he was supposed to have received from Galileo, who, during the whole time of the appearance of these comets, was confined to his bed by severe illness. This essay was printed in Florence *at the sign of The Medicean Stars.** What principally deserves notice in it, is the opinion of Galileo, that the distance of a comet cannot be safely determined by its parallax, from which we learn that he inclined to believe that comets are nothing but meteors occasionally appearing in the atmosphere, like rainbows, parhelia, and similar phenomena. He points out the difference in this respect between a fixed object, the distance of which may be calculated from the difference of direction in which two observers (at a known distance from each other) are obliged to turn themselves in order to see it, and meteors like the rainbow, which are simultaneously formed in different drops of water for each spectator, so that two

---

* In Firenze nella Stamperia di Pietro Cecconcelli alle stelle Medicee, 1619.

E 2

observers in different places are in fact contemplating different objects. He then warns astronomers not to engage with too much warmth in a discussion on the distance of comets before they assure themselves to which of these two classes of phenomena they are to be referred. The remark is in itself perfectly just, although the opinion which occasioned it is now as certainly known to be, erroneous, but it is questionable whether the observations which, up to that time, had been made upon comets, were sufficient, either in number or quality, to justify the censure which has been cast on Galileo for his opinion. The theory, moreover, is merely introduced as an hypothesis in Guiducci's essay. The same opinion was for a short time embraced by Cassini, a celebrated Italian astronomer, invited by Louis XIV. to the Observatory at Paris, when the science was considerably more advanced, and Newton, in his *Principia*, did not think it unworthy of him to show on what grounds it is untenable.

Galileo was become the object of animosity in so many quarters that none of his published opinions, whether correct or incorrect, ever wanted a ready antagonist. The champion on the present occasion was again a Jesuit; his name was Oratio Grassi, who published *The Astronomical and Philosophical Balance,* under the disguised signature of Lotario Sarsi.

Galileo and his friends were anxious that his reply to Grassi should appear as quickly as possible, but his health had become so precarious and his frequent illnesses occasioned so many interruptions, that it was not until the autumu of 1623 that Il Saggiatore (or The Assayer) as he called his answer, was ready for publication. This was printed by the Lyncean Academy, and as Cardinal Maffeo Barberino, who had just been elected Pope, (with the title of Urban VIII.) had been closely connected with that society, and was also a personal friend of Cesi, and of Galileo, it was thought a prudent precaution to dedicate the pamphlet to him. This essay enjoys a peculiar reputation among Galileo's works, not only for the matter contained in it, but also for the style in which it is written; insomuch that Andrès*, when eulogizing Galileo as one of the earliest who adorned philosophical truths with the graces and ornaments of language, expressly instances the Saggiatore, which is also quoted by Frisi and Algarotti, as a perfect model of this sort of composition. In the latter particular, it is unsafe to interfere with the decisions of an Italian critic; but with respect to its substance, this famous composition scarcely appears to deserve its preeminent reputation. It is a prolix and rather tedious examination of Grassi's Essay; nor do the arguments seem so satisfactory, nor the reasonings so compact as is generally the case in Galileo's other writings. It does however, like all his other works, contain many very remarkable passages, and the celebrity of this production requires that we should extract one or two of the most characteristic.

The first, though a very short one, will serve to shew the tone which Galileo had taken with respect to the Copernican system since its condemnation at Rome, in 1616. " In conclusion, since the motion attributed to the earth, which I, as a pious and Catholic person, consider most false, and not to exist, accommodates itself so well to explain so many and such different phenomena, I shall not feel sure, unless Sarsi descends to more distinct considerations than those which he has yet produced, that, false as it is, it may not just as deludingly correspond with the phenomena of comets."

Sarsi had quoted a story from Suidas in support of his argument that motion always produces heat, how the Babylomans used to cook their eggs by whirling them in a sling; to which Galileo replies: " I cannot refrain from marvelling that Sarsi will persist in proving to me, by authorities, that which at any moment I can bring to the test of experiment. We examine witnesses in things which are doubtful, past, and not permanent, but not in those things which are done in our own presence. If discussing a difficult problem were like carrying a weight, since several horses will carry more sacks of corn than one alone will, I would agree that many reasoners avail more than one; but *discoursing* is like *coursing*, and not like carrying, and one barb by himself will run farther than a hundred Friesland horses. When Sarsi brings up such a multitude of authors, it does not seem to me that he in the least degree strengthens his own conclusions, but he ennobles the cause of Signor Mario and myself, by shewing that we reason better than many men of established reputation. If Sarsi insists that I believe,

---

* Dell' Origine d'ogni Literatura: Parma, 1787.

on Suidas' credit, that the Babylonians cooked eggs by swiftly whirling them in a sling, I will believe it; but I must needs say, that the cause of such an effect is very remote from that to which it is attributed, and to find the true cause I shall reason thus. If an effect does not follow with us which followed with others at another time, it is because, in our experiment, something is wanting which was the cause of the former success; and if only one thing is wanting to us, that one thing is the true cause. Now we have eggs, and slings, and strong men to whirl them, and yet they will not become cooked; nay, if they were hot at first, they more quickly become cold: and since nothing is wanting to us but to be Babylonians, it follows that being Babylonians is the true cause why the eggs became hard, and not the friction of the air, which is what I wished to prove.—Is it possible that in travelling post, Sarsi has never noticed what freshness is occasioned on the face by the continual change of air? and if he has felt it, will he rather trust the relation by others, of what was done two thousand years ago at Babylon, than what he can at this moment verify in his own person? I at least will not be so wilfully wrong, and so ungrateful to nature and to God, that having been gifted with sense and language, I should voluntarily set less value on such great endowments than on the fallacies of a fellow man, and blindly and blunderingly believe whatever I hear, and barter the freedom of my intellect for slavery to one as liable to error as myself."

Our final extract shall exhibit a sample of Galileo's metaphysics, in which may be observed the germ of a theory very closely allied to that which was afterwards developed by Locke and Berkeley.—" I have now only to fulfil my promise of declaring my opinions on the proposition that motion is the cause of heat, and to explain in what manner it appears to me that it may be true. But I must first make some remarks on that which we call heat, since I strongly suspect that a notion of it prevails which is very remote from the truth; for it is believed that there is a true accident, affection, and quality, really inherent in the substance by which we feel ourselves heated. This much I have to say, that so soon as I conceive a material or corporeal substance, I simultaneously feel the necessity of conceiving that it has its boundaries, and is of some shape or other; that, relatively to others, it is great or small; that it is in this or that place, in this or that time; that it is in motion, or at rest; that it touches, or does not touch another body; that it is unique, rare, or common; nor can I, by any act of the imagination, disjoin it from these qualities: but I do not find myself absolutely compelled to apprehend it as necessarily accompanied by such conditions, as that it must be white or red, bitter or sweet, sonorous or silent, smelling sweetly or disagreeably; and if the senses had not pointed out these qualities, it is probable that language and imagination alone could never have arrived at them. Because, I am inclined to think that these tastes, smells, colours, &c., with regard to the subject in which they appear to reside, are nothing more than mere names, and exist only in the sensitive body; insomuch that, when the living creature is removed, all these qualities are carried off and annihilated; although we have imposed particular names upon them, and different from those of the other first and real accidents, and would fain persuade ourselves that they are truly and in fact distinct. But I do not believe that there exists any thing in external bodies for exciting tastes, smells, and sounds, but size, shape, quantity, and motion, swift or slow; and if ears, tongues, and noses were removed, I am of opinion that shape, number, and motion would remain, but there would be an end of smells, tastes, and sounds, which, abstractedly from the living creature, I take to be mere words."

In the spring following the publication of the "Saggiatore," that is to say, about the time of Easter, in 1624, Galileo went a third time to Rome to compliment Urban on his elevation to the pontifical chair. He was obliged to make this journey in a litter; and it appears from his letters that for some years he had been seldom able to bear any other mode of conveyance. In such a state of health it seems unlikely that he would have quitted home on a mere visit of ceremony, which suspicion is strengthened by the beginning of a letter from him to Prince Cesi, dated in October, 1623, in which he says: " I have received the very courteous and prudent advice of your excellency about the time and manner of my going to Rome, and shall act upon it; and I will visit you at Acqua Sparta, that I may be completely informed of the actual state of things at Rome." However this may be, nothing could be more gratifying than his public reception there. His stay in Rome did not exceed two months, (from the beginning of April till June,) and during that time he was admitted to six long and satisfactory interviews with the Pope, and on his departure received the promise of a pension for his son Vincenzo, and was himself presented with " a fine painting, two medals, one of gold and the other of silver, and a good quantity of agnus dei." He had also much communication with several of the cardinals, one of whom, Cardinal Hohenzoller, told him that he had represented to the pope on the subject of Copernicus, that " all the heretics were of that opinion, and considered it as undoubted; and that it would be necessary to be very circumspect in coming to any resolution: to which his holiness replied, that the church had not condemned it, nor was it to be condemned as heretical, but only as rash ; adding, that there was no fear of any one undertaking to prove that it must necessarily be true." Urban also addressed a letter to Ferdinand, who had succeeded his father Cosmo as Grand Duke of Tuscany, expressly for the purpose of recommending Galileo to him. " For We find in him not only literary distinction, but also the love of piety, and he is strong in those qualities by which pontifical good-will is easily obtained. And now, when he has been brought to this city to congratulate Us on Our elevation, We have very lovingly embraced him ;—nor can We suffer him to return to the country whither your liberality recalls him without an ample provision of pontifical love. And that you may know how dear he is to Us, We have willed to give him this honourable testimonial of virtue and piety. And We further signify that every benefit, which you shall confer upon him, imitating, or even surpassing your father's liberality, will conduce to Our gratification." Honoured with these unequivocal marks of approbation, Galileo returned to Florence.

His son Vincenzo is soon afterwards spoken of as being at Rome ; and it is not improbable that Galileo sent him thither on the appointment of his friend and pupil, the Abbé Castelli, to be mathematician to the pope. Vincenzo had been legitimated by an edict of Cosmo in 1619, and, according to Nelli, married, in 1624, Sestilia, the daughter of Carlo Bocchineri. There are no traces to be found of Vincenzo's mother after 1610, and perhaps she died about that time. Galileo's family by her consisted of Vincenzo and two daughters, Julia and Polissena, who both took the veil in the convent of Saint Matthew at Arcetri, under the names of Sister Arcangiola and Sister Maria Celeste. The latter is said to have possessed extraordinary talents. The date of Vincenzo's marriage, as given by Nelli, appears somewhat inconsistent with the correspondence between Galileo and Castelli, in which, so late as 1629, Galileo is apparently writing of his son as a student under Castelli's superintendence, and intimates the amount of pocket-money he can afford to allow him, which he fixes at three crowns a month; adding, that "he ought to be contented with as many crowns, as, at his age, I possessed groats." Castelli had given but an unfavourable account of Vincenzo's conduct, characterizing him as "dissolute, obstinate, and impudent;" in consequence of which behaviour, Galileo seems to have thought that the pension of sixty crowns, which had been granted by the pope, might be turned to better account than by employing it on his son's education ; and accordingly in his reply he requested Castelli to dispose of it, observing that the proceeds would be useful in assisting him to discharge a great load of debt with which he found himself saddled on account of his brother's family. Besides this pension, another of one hundred crowns was in a few years granted by Urban to Galileo himself, but it appears to have been very irregularly paid, if at all.

About the same time Galileo found himself menaced either with the deprivation of his stipend as extraordinary professor at Pisa, or with the loss of that leisure which, on his removal to Florence, he had been so anxious to secure. In 1629, the question was agitated by the party opposed to him, whether it were in the power of the grand duke to assign a pension out of the funds of the University, arising out of ecclesiastical dues, to one who neither lectured nor resided there. This scruple had slept during nineteen years which had elapsed since Galileo's establishment in Florence, but probably those who now raised it reckoned upon finding in Ferdinand II., then scarcely of age, a less firm supporter of Galileo than his father Cosmo had been. But the matter did not proceed so far; for, after full deliberation, the prevalent opinion of the theologians and jurists who were consulted appeared to be in favour of this exercise of prerogative, and accordingly Galileo retained his stipend and privileges.

CHAPTER XIII.

Publication of Galileo's ' System of the World'—His Condemnation and Abjuration.

IN the year 1630, Galileo brought to its conclusion his great work, "The Dialogue on the Ptolemaic and Copernican Systems," and began to take the necessary steps for procuring permission to print it. This was to be obtained in the first instance from an officer at Rome, entitled the master of the sacred palace; and after a little negotiation Galileo found it would be necessary for him again to return thither, as his enemies were still busy in thwarting his views and wishes. Niccolo Riccardi, who at that time filled the office of master of the palace, had been a pupil of Galileo, and was well disposed to facilitate his plans; he pointed out, however, some expressions in the work which he thought it necessary to erase, and, with the understanding that this should be done, he returned the manuscript to Galileo with his subscribed approbation. The unhealthy season was drawing near, and Galileo, unwilling to face it, returned home, where he intended to complete the index and dedication, and then to send it back to Rome to be printed in that city, under the superintendence of Federigo Cesi. This plan was disconcerted by the premature death of that accomplished nobleman, in August 1630, in whom Galileo lost one of his steadiest and most effective friends and protectors. This unfortunate event determined Galileo to attempt to procure permission to print his book at Florence. A contagious disorder had broken out in Tuscany with such severity as almost to interrupt all communication between Florence and Rome, and this was urged by Galileo as an additional reason for granting his request. Riccardi at first seemed inclined to insist that the book should be sent to him a second time, but at last contented himself with inspecting the commencement and conclusion, and consented that (on its receiving also a license from the inquisitor-general at Florence, and from one or two others whose names appear on the title-page) it might be printed where Galileo wished.

These protracted negotiations prevented the publication of the work till late in 1632; it then appeared, with a dedication to Ferdinand, under the following title:—" A Dialogue, by Galileo Galilei, Extraordinary Mathematician of the University of Pisa, and Principal Philosopher and Mathematician of the Most Serene Grand Duke of Tuscany; in which, in a conversation of four days, are discussed the two principal Systems of the World, the Ptolemaic and Copernican, indeterminately proposing the Philosophical Arguments as well on one side as on the other." The beginning of the introduction, which is addressed "To the discreet Reader," is much too characteristic to be passed by without notice.—" Some years ago, a salutary edict was promulgated at Rome, which, in order to obviate the perilous scandals of the present age, enjoined an opportune silence on the Pythagorean opinion of the earth's motion. Some were not wanting, who rashly asserted that this decree originated, not in a judicious examination, but in ill informed passion; and complaints were heard that counsellors totally inexperienced in astronomical observations ought not by hasty prohibitions to clip the wings of speculative minds. My zeal could not keep silence when I heard these rash lamentations, and I thought it proper, as being fully informed with regard to that most prudent determination, to appear publicly on the theatre of the world as a witness of the actual truth. I happened at that time to be in Rome: I was admitted to the audiences, and enjoyed the approbation of the most eminent prelates of that court, nor did the publication of that decree occur without my receiving some prior intimation of it.* Wherefore it is my intention in this present work, to show to foreign nations that as much is known of this matter in Italy, and particularly in Rome, as ultramontane diligence can ever have formed any notion of, and collecting together all my own speculations on the Copernican system, to give them to understand that the knowledge of all these preceded the Roman censures, and that from this

---

* Delambre quotes this sentence from a passage which is so obviously ironical throughout, as an instance of Galileo's mis-statement of facts!—*Hist. de l'Astr. Mod.*, vol. i. p. 666.

country proceed not only dogmas for the salvation of the soul, but also ingenious discoveries for the gratification of the understanding. With this object, I have taken up in the Dialogue the Copernican side of the question, treating it as a pure mathematical hypothesis; and endeavouring in every artificial manner to represent it as having the advantage, not over the opinion of the stability of the earth absolutely, but according to the manner in which that opinion is defended by some, who indeed profess to be Peripatetics, but retain only the name, and are contented without improvement to worship shadows, not philosophizing with their own reason, but only from the recollection of four principles imperfectly understood."—This very flimsy veil could scarcely blind any one as to Galileo's real views in composing this work, nor does it seem probable that he framed it with any expectation of appearing neutral in the discussion. It is more likely that he flattered himself that, under the new government at Rome, he was not likely to be molested on account of the personal prohibition which he had received in 1616, "not to believe or teach the motion of the earth in any manner," provided he kept himself within the letter of the limits of the more public and general order, that the Copernican system was not to be brought forward otherwise than as a mere mathematically convenient, but in fact unreal supposition. So long as this decree remained in force, a due regard to consistency would compel the Roman Inquisitors to notice an unequivocal violation of it; and this is probably what Urban had implied in the remark quoted by Hohenzoller to Galileo.* There were not wanting circumstances which might compensate for the loss of Cosmo and of Federigo Cesi; Cosmo had been succeeded by his son, who, though he had not yet attained his father's energy, showed himself as friendly as possible to Galileo. Cardinal Bellarmine, who had been mainly instrumental in procuring the decree of 1616, was dead; Urban on the contrary, who had been among the few Cardinals who then opposed it as uncalled for and ill-advised, was now possessed of supreme power, and his recent affability seemed to prove that the increased difference in their stations had not caused him to forget their early and long-continued intimacy. It is probable that Galileo would not have found himself mistaken in this estimate of his position, but for an unlucky circumstance, of which his enemies immediately saw the importance, and which they were not slow in making available against him. The dialogue of Galileo's work is conducted between three personages;—Salviati and Sagredo, who were two noblemen, friends of Galileo, and Simplicio, a name borrowed from a noted commentator upon Aristotle, who wrote in the sixth century. Salviati is the principal philosopher of the work; it is to him that the others apply for solutions of their doubts and difficulties, and on him the principal task falls of explaining the tenets of the Copernican theory. Sagredo is only a half convert, but an acute and ingenious one; to him are allotted the objections which seem to have some real difficulty in them, as well as lively illustrations and digressions, which might have been thought inconsistent with the gravity of Salviati's character. Simplicio, though candid and modest, is of course a confirmed Ptolemaist and Aristotelian, and is made to produce successively all the popular arguments of that school in support of his master's system. Placed between the wit and the philosopher, it may be guessed that his success is very indifferent, and in fact he is alternately ridiculed and confuted at every turn. As Galileo racked his memory and invention to leave unanswered no argument which was or could be advanced against Copernicus, it unfortunately happened, that he introduced some which Urban himself had urged upon him in their former controversies on this subject; and Galileo's opponents found means to make His Holiness believe that the character of Simplicio had been sketched in personal derision of him. We do not think it necessary to exonerate Galileo from this charge; the obvious folly of such an useless piece of ingratitude speaks sufficiently for itself. But self-love is easily irritated; and Urban, who aspired to a reputation for literature and science, was peculiarly sensitive on this point. His own expressions almost prove his belief that such had been Galileo's design, and it seems to explain the otherwise inexplicable change which took place in his conduct towards his old friend, on account of a book which he had himself undertaken to examine, and of which he had authorised the publication.

One of the earliest notices of what was approaching, is found in the dispatches,

---

* Page 54.

dated August 24, 1632, from Ferdinand's minister, Andrea Cioli, to Francesco Nicolini, the Tuscan ambassador at the court of Rome. "I have orders to signify to Your Excellency that His Highness remains greatly astonished that a book, placed by the author himself in the hands of the supreme authority in Rome, read and read again there most attentively, and in which every thing, not only with the consent, but at the request of the author, was amended, altered, added, or removed at the will of his superiors, which was again subjected here to the same examination, agreeably to orders from Rome, and which finally was licensed both there and here, and here printed and published, should now become an object of suspicion at the end of two years, and the author and printer be prohibited from publishing any more."—In the sequel is intimated Ferdinand's desire that the charges, of whatever nature they might be, either against Galileo or his book, might be reduced to writing and forwarded to Florence, that he might prepare for his justification; but this reasonable demand was utterly disregarded. It appears to have been owing to the mean subserviency of Cioli to the court of Rome, that Ferdinand refrained from interfering more strenuously to protect Galileo. Cioli's words are: "The Grand Duke is so enraged with this business of Galileo, that I do not know what will be done. I know, at least, that His Holiness shall have no reason to complain of his ministers, or of their bad advice."*

A letter from Galileo's Venetian friend Micanzio, dated about a month later, is in rather a bolder and less formal style:—"The efforts of your enemies to get your book prohibited will occasion no loss either to your reputation, or to the intelligent part of the world. As to posterity, this is just one of the surest ways to hand the book down to them. But what a wretched set this must be to whom every good thing, and all that is founded in nature, necessarily appears hostile and odious! The world is not restricted to a single corner; you will see the book printed in more places and languages than one; and just for this reason, I wish they would prohibit all good books. My disgust arises from seeing myself deprived of what I most desire of this sort, I mean your other dialogues; and if, from this cause, I fail in having the pleasure of seeing them, I shall devote to a hundred thousand devils these unnatural and godless hypocrites."

At the same time, Thomas Campanella, a monk, who had already distinguished himself by an apology for Galileo (published in 1622), wrote to him from Rome:—" I learn with the greatest disgust, that a congregation of angry theologians is forming to condemn your Dialogues, and that no single member of it has any knowledge of mathematics, or familiarity with abstruse speculations. I should advise you to procure a request from the Grand Duke that, among the Dominicans and Jesuits and Theatins, and secular priests whom they are putting on this congregation against your book, they should admit also Castelli and myself." It appears, from subsequent letters both from Campanella and Castelli, that the required letter was procured and sent to Rome, but it was not thought prudent to irritate the opposite party by a request which it was then clearly seen would have been made in vain. Not only were these friends of Galileo, not admitted to the congregation, but, upon some pretext, Castelli was even sent away from Rome, as if Galileo's enemies desired to have as few enlightened witnesses as possible of their proceedings; and on the contrary, Scipio Chiaramonte, who had been long known for one of the staunchest and most bigoted defenders of the old system, and who, as Montucla says, seems to have spent a long life in nothing but retarding, as far as he was able, the progress of discovery, was summoned from Pisa to complete their number. From this period we have a tolerably continuous account of the proceedings against Galileo, in the dispatches which Nicolini sent regularly to his court. It appears from them that Nicolini had several interviews with the Pope, whom he found highly incensed against Galileo, and in one of the earliest he received an intimation to advise the Duke " not to engage himself in this matter as he had done in the other business of Alidosi,* because he would not get through it with honour." Finding Urban in this humour, Nicolini thought it best to temporize, and to avoid the appearance of any thing like direct opposition. On the 15th of September, probably as soon as the first report on

* Alidosi was a Florentine nobleman, whose estate Urban wished to confiscate on a charge of heresy.— *Galuzzi.*

* Galuzzi, Storia di Toscana. Firenze, 1822.

Galileo's book had been made, Nicolini received a private notice from the Pope, "in especial token of the esteem in which he held the Grand Duke," that he was unable to do less than consign the work to the consideration of the Inquisition. Nicolini was permitted to communicate this to the Grand Duke only, and both were declared liable to "the usual censures" of the Inquisition in case of divulging the secret.

The next step was to summon Galileo to Rome, and the only answer returned to all Nicolini's representations of his advanced age of seventy years, the very infirm state of his health, and the discomforts which he must necessarily suffer in such a journey, and in keeping quarantine, was that he might come at leisure, and that the quarantine should be relaxed as much as possible in his favour, but that it was indispensably necessary that he should be personally examined before the Inquisition at Rome. Accordingly, on the 14th of February, 1633, Nicolini announces Galileo's arrival, and that he had officially notified his presence to the Assessor and Commissary of the Holy Office. Cardinal Barberino, Urban's nephew, who seems on the whole to have acted a friendly part towards Galileo, intimated to him that his most prudent course would be to keep himself as much at home and as quiet as possible, and to refuse to see any but his most intimate friends. With this advice, which was repeated to him from several quarters, Galileo thought it best to comply, and kept himself entirely secluded in Nicolini's palace, where he was as usual maintained at the expense of the Grand Duke. Nelli quotes two letters, which passed between Ferdinand's minister Cioli and Nicolini, in which the former intimated that Galileo's expenses were to be defrayed only during the first month of his residence at Rome. Nicolini returned a spirited answer, that in that case, after the time specified, he should continue to treat him as before at his own private cost.

The permission to reside at the ambassador's palace whilst his cause was pending, was granted and received as an extraordinary indulgence on the part of the Inquisition, and indeed if we estimate the proceedings throughout against Galileo by the usual practice of that detestable tribunal, it will appear that he was treated with unusual consideration. Even when it became necessary in the course of the inquiry to examine him in person, which was in the beginning of April, although his removal to the Holy Office was then insisted upon, yet he was not committed to close or strictly solitary confinement. On the contrary, he was honourably lodged in the apartments of the Fiscal of the Inquisition, where he was allowed the attendance of his own servant, who was also permitted to sleep in an adjoining room, and to come and go at pleasure. His table was still furnished by Nicolini. But, notwithstanding the distinction with which he was thus treated, Galileo was annoyed and uneasy at being (though little more than nominally) within the walls of the Inquisition. He became exceedingly anxious that the matter should be brought to a conclusion, and a severe attack of his constitutional complaints rendered him still more fretful and impatient. On the last day of April, about ten days after his first examination, he was unexpectedly permitted to return to Nicolini's house, although the proceedings were yet far from being brought to a conclusion. Nicolini attributes this favour to Cardinal Barberino, who, he says, liberated Galileo on his own responsibility, in consideration of the enfeebled state of his health.

In the society of Nicolini and his family, Galileo recovered something of his courage and ordinary cheerfulness, although his return appears to have been permitted on express condition of a strict seclusion; for at the latter end of May, Nicolini was obliged to apply for permission that Galileo should take that exercise in the open air which was necessary for his health; on which occasion he was permitted to go into the public gardens in a half-closed carriage.

On the evening of the 20th of June, rather more than four months after Galileo's arrival in Rome, he was again summoned to the Holy Office, whither he went the following morning; he was detained there during the whole of that day, and on the next day was conducted in a penitential dress * to the Convent of Minerva, where the Cardinals and Prelates, his judges, were assembled for the purpose of passing judgment upon him, by which this venerable old man was solemnly called upon to renounce and abjure, as impious and heretical, the opinions which his whole existence had been consecrated to form and strengthen.

---

* S' irrito il Papa, e lo fece abjurare, comparendo il pover uomo con uno straccio di camicia indosso, che faceva compassione, MS. nella Bibl. Magliab. Venturi.

As we are not aware that this remarkable record of intolerance and bigoted folly has ever been printed entire in English, we subjoin a literal translation of the whole sentence and abjuration.

The Sentence of the Inquisition on Galileo.

"We, the undersigned, by the Grace of God, Cardinals of the Holy Roman Church, Inquisitors General throughout the whole Christian Republic, Special Deputies of the Holy Apostolical Chair against heretical depravity,

"Whereas you, Galileo, son of the late Vincenzo Galilei of Florence, aged seventy years, were denounced in 1615 to this Holy Office, for holding as true a false doctrine taught by many, namely, that the sun is immoveable in the centre of the world, and that the earth moves, and also with a diurnal motion; also, for having pupils whom you instructed in the same opinions ; also, for maintaining a correspondence on the same with some German mathematicians ; also for publishing certain letters on the solar spots, in which you developed the same doctrine as true; also, for answering the objections which were continually produced from the Holy Scriptures, by glozing the said Scriptures according to your own meaning; and whereas thereupon was produced the copy of a writing, in form of a letter, professedly written by you to a person formerly your pupil, in which, following the hypotheses of Copernicus, you include several propositions contrary to the true sense and authority of the Holy Scripture : therefore this holy tribunal being desirous of providing against the disorder and mischief which was thence proceeding and increasing to the detriment of the holy faith, by the desire of His Holiness, and of the Most Eminent Lords Cardinals of this supreme and universal Inquisition, the two propositions of the stability of the sun, and motion of the earth, were *qualified* by the *Theological Qualifiers* as follows :

"1st. *The proposition that the Sun is in the centre of the world and immoveable from its place, is absurd, philosophically false, and formally heretical ; because it is expressly contrary to the Holy Scripture.*

"2dly. *The proposition that the Earth is not the centre of the world, nor immoveable, but that it moves, and also with a diurnal motion, is also absurd, philosophically false, and, theologically considered, at least erroneous in faith.*

"But whereas being pleased at that time to deal mildly with you, it was decreed in the Holy Congregation, held before His Holiness on the 25th day of February, 1616, that His Eminence the Lord Cardinal Bellarmine should enjoin you to give up altogether the said false doctrine; if you should refuse, that you should be ordered by the Commissary of the Holy Office to relinquish it, not to teach it to others, nor to defend it, nor ever mention it, and in default of acquiescence that you should be imprisoned; and in execution of this decree, on the following day at the palace, in presence of His Eminence the said Lord Cardinal Bellarmine, after you had been mildly admonished by the said Lord Cardinal, you were commanded by the acting Commissary of the Holy Office, before a notary and witnesses, to relinquish altogether the said false opinion, and in future neither to defend nor teach it in any manner, neither verbally nor in writing, and upon your promising obedience you were dismissed.

"And in order that so pernicious a doctrine might be altogether rooted out, nor insinuate itself farther to the heavy detriment of the Catholic truth, a decree emanated from the Holy Congregation of the Index* prohibiting the books which treat of this doctrine; and it was declared false, and altogether contrary to the Holy and Divine Scripture.

"And whereas a book has since appeared, published at Florence last year, the title of which shewed that you were the author, which title is : *The Dialogue of Galileo Galilei, on the two principal systems of the world, the Ptolemaic and Copernican ;* and whereas the Holy Congregation has heard that, in consequence of the printing of the said book, the false opinion of the earth's motion and stability of the sun is daily gaining ground ; the said book has been taken into careful consideration, and in it has been detected a glaring violation of the said order, which had been intimated to you ; inasmuch as in this book you have

---

* The Index is a list of books, the reading of which is prohibited to Roman Catholics. This list, in the early periods of the Reformation, was often consulted by the curious, who were enlarging their libraries ; and a story is current in England, that, to prevent this mischief, the Index itself was inserted in its own forbidden catalogue. The origin of this story is, that an Index was published in Spain, particularizing the objectionable passages in such books as were only partially condemned ; and although compiled with the best intentions, this was found to be so racy, that it became necessary to forbid the circulation of this edition in subsequent lists.

defended the said opinion, already and in your presence condemned; although in the said book you labour with many circumlocutions to induce the belief that it is left by you undecided, and in express terms probable: which is equally a very grave error, since an opinion can in no way be probable which has been already declared and finally determined contrary to the divine Scripture. Therefore by Our order you have been cited to this Holy Office, where, on your examination upon oath, you have acknowledged the said book as written and printed by you. You also confessed that you began to write the said book ten or twelve years ago, after the order aforesaid had been given. Also, that you demanded license to publish it, but without signifying to those who granted you this permission that you had been commanded not to hold, defend, or teach the said doctrine in any manner. You also confessed that the style of the said book was, in many places, so composed that the reader might think the arguments adduced on the false side to be so worded as more effectually to entangle the understanding than to be easily solved, alleging in excuse, that you have thus run into an error, foreign (as you say) to your intention, from writing in the form of a dialogue, and in consequence of the natural complacency which every one feels with regard to his own subtilties, and in showing himself more skilful than the generality of mankind in contriving, even in favour of false propositions, ingenious and apparently probable arguments.

"And, upon a convenient time being given to you for making your defence, you produced a certificate in the handwriting of His Eminence the Lord Cardinal Bellarmine, procured, as you said, by yourself, that you might defend yourself against the calumnies of your enemies, who reported that you had abjured your opinions, and had been punished by the Holy Office; in which certificate it is declared, that you had not abjured, nor had been punished, but merely that the declaration made by His Holiness, and promulgated by the Holy Congregation of the Index, had been announced to you, which declares that the opinion of the motion of the earth, and stability of the sun, is contrary to the Holy Scriptures, and, therefore, cannot be held or defended. Wherefore, since no mention is there made of two articles of the order, to wit, the order 'not to teach,' and 'in any manner,' you argued that we ought to believe that, in the lapse of fourteen or sixteen years, they had escaped your memory, and that this was also the reason why you were silent as to the order, when you sought permission to publish your book, and that this is said by you not to excuse your error, but that it may be attributed to vain-glorious ambition, rather than to malice. But this very certificate, produced on your behalf, has greatly aggravated your offence, since it is therein declared that the said opinion is contrary to the Holy Scripture, and yet you have dared to treat of it, to defend it, and to argue that it is probable; nor is there any extenuation in the licence artfully and cunningly extorted by you, since you did not intimate the command imposed upon you. But whereas it appeared to Us that you had not disclosed the whole truth with regard to your intentions, We thought it necessary to proceed to the rigorous examination of you, in which (without any prejudice to what you had confessed, and which is above detailed against you, with regard to your said intention) you answered like a good Catholic.

"Therefore, having seen and maturely considered the merits of your cause, with your said confessions and excuses, and every thing else which ought to be seen and considered, We have come to the underwritten final sentence against you.

"Invoking, therefore, the most holy name of Our Lord Jesus Christ, and of His Most Glorious Virgin Mother Mary, by this Our final sentence, which, sitting in council and judgment for the tribunal of the Reverend Masters of Sacred Theology, and Doctors of both Laws, Our Assessors, We put forth in this writing touching the matters and controversies before Us, between The Magnificent Charles Sincerus, Doctor of both Laws, Fiscal Proctor of this Holy Office of the one part, and you, Galileo Galilei, an examined and confessed criminal from this present writing now in progress as above of the other part, We pronounce, judge, and declare, that you, the said Galileo, by reason of these things which have been detailed in the course of this writing, and which, as above, you have confessed, have rendered yourself vehemently suspected by this Holy Office of heresy: that is to say, that you believe and hold the false doctrine, and contrary to the Holy and Divine Scriptures, namely, that the sun is the centre of the world, and that it does not move from east to west, and that the earth does move, and is not the centre of the world; also that an opinion can be held and supported as probable after 'it has been declared and finally decreed contrary to the Holy Scripture, and consequently that you have incurred all the censures and penalties enjoined and promulgated in the sacred canons, and other general and particular constitutions against delinquents of this description. From which it is Our pleasure that you be absolved, provided that, first, with a sincere heart and unfeigned faith, in Our presence, you abjure, curse, and detest the said errors and heresies, and every other error and heresy contrary to the Catholic and Apostolic Church of Rome, in the form now shown to you.

"But, that your grievous and pernicious error and transgression may not go altogether unpunished, and that you may be made more cautious in future, and may be a warning to others to abstain from delinquencies of this sort, We decree that the book of the dialogues of Galileo Galilei be prohibited by a public edict, and We condemn you to the formal prison of this Holy Office for a period determinable at Our pleasure; and, by way of salutary penance, We order you, during the next three years, to recite once a week the seven penitential psalms, reserving to Ourselves the power of moderating, commuting, or taking off the whole or part of the said punishment and penance.

"And so We say, pronounce, and by Our sentence declare, decree, and reserve, in this and in every other better form and manner, which lawfully We may and can use.

"So We, the subscribing Cardinals, pronounce.

Felix, Cardinal di Ascoli,
Guido, Cardinal Bentivoglio,
Desiderio, Cardinal di Cremona,
Antonio, Cardinal S. Onofrio,
Berlingero, Cardinal Gessi,
Fabricio, Cardinal Verospi,
Martino, Cardinal Ginetti."

We cannot suppose that Galileo, even broken down as he was with age and infirmities, and overawed by the merciless tribunal to whose power he was subjected, could without extreme reluctance thus formally give the lie to his whole life, and call upon God to witness his renunciation of the opinions which even his bigoted judges must have felt that he still clung to in his heart."

We know indeed that his friends were unanimous in recommending an unqualified acquiescence in whatever might be required, but some persons have not been able to find an adequate explanation of his submission, either in their exhortations, or in the mere dread of the alternative which might await him in case of non-compliance. It has in short been, supposed, although the suspicion scarcely rests upon grounds sufficiently strong to warrant the assertion, that Galileo did not submit to this abjuration until forced to it, not merely by the apprehension, but by the actual experience of personal violence. The arguments on which this horrible idea appears to be mainly founded are the two following: First, the Inquisitors declare in their sentence that, not satisfied with Galileo's first confession, they judged it necessary to proceed " to the rigorous examination of him, in which he answered like a good Catholic.*" It is pretended by those who are more familiar with inquisitorial language than we can profess to be, that the words *il rigoroso esame*, form the official phrase for the application of the torture, and accordingly they interpret this passage to mean, that the desired answers and submission had thus been extorted from Galileo, which his judges had otherwise failed to get from him. And, secondly, the partisans of this opinion bring forward in corroboration of it, that Galileo immediately on his departure from Rome, in addition to his old complaints, was found to be afflicted with hernia, and this was a common consequence of the torture of the cord, which they suppose to have been inflicted. It is right to mention that no other trace can be found of this supposed torturing in all the documents relative to the proceedings against Galileo, at least Venturi was so assured by one who had inspected the originals at Paris.†

---

* Giudicassimo esser necessario Venir contro di te al rigoroso esame nel quale rispondesti cattolicamente.

† The fate of these documents is curious; after being long preserved at Rome, they were carried away in 1809, by order of Buonaparte, to Paris, where they remained till his first abdication. Just before the hundred days, the late king of France, wishing to inspect them, ordered that they should be brought to his own apartments for that purpose. In the hasty flight which soon afterwards followed, the manuscripts were forgotten, and it is not known what became of them. A French translation, begun by Napoleon's desire, was completed only down to the 30th of April, 1633, the date of Galileo's first return to Nicolini's palace.

Although the arguments we have mentioned appear to us slight, yet neither can we attach much importance to the contrast which the favourers of the opposite opinion profess to consider so incredible between the honourable manner in which Galileo was treated throughout the rest of the inquiry, and the suspected harsh proceeding against him. Whether Galileo should be lodged in a prison or a palace, was a matter of far other importance to the Inquisitors and to their hold upon public opinion, than the question whether or not he should be suffered to exhibit a persevering resistance to the censures which they were prepared to cast upon him. Nor need we shrink from the idea, as we might from suspecting of some gross crime, on trivial grounds, one of hitherto unblemished innocence and character. The question may be disencumbered of all such scruples, since one atrocity more or less can do little towards affecting our judgment of the unholy Office of the Inquisition.

Delambre, who could find so much to reprehend in Galileo's former uncompromising boldness, is deeply penetrated with the insincerity of his behaviour on the present occasion. He seems to have forgotten that a tribunal which finds it convenient to carry on its inquiries in secret, is always liable to the suspicion of putting words into the mouth of its victims; and if it were worth while, there is sufficient internal evidence that the language which Galileo is made to hold in his defence and confession, is rather to be read as the composition of his judges than his own. For instance, in one of the letters which we have extracted*, it may be seen that this obnoxious work was already in forward preparation as early as 1610, and yet he is made to confess, and the circumstance appears to be brought forward in aggravation of his guilt, that he began to write it after the prohibition which he had received in 1616.

The abjuration was drawn up in the following terms:—

The Abjuration of Galileo.

" I Galileo Galilei, son of the late Vincenzo Galilei, of Florence, aged 70 years, being brought personally to judgment, and kneeling before you, Most Eminent and Most Reverend Lords Cardinals, General Inquisitors of the universal Christian republic against heretical depravity, having before my eyes the Holy Gospels, which I touch with my own hands, swear, that I have always believed, and now believe, and with the help of God will in future believe, every article which the Holy Catholic and Apostolic Church of Rome holds, teaches, and preaches. But because I had been enjoined by this Holy Office altogether to abandon the false opinion which maintains that the sun is the centre and immoveable, and forbidden to hold, defend, or teach, the said false doctrine in any manner, and after it had been signified to me that the said doctrine is repugnant with the Holy Scripture, I have written and printed a book, in which I treat of the same doctrine now condemned, and adduce reasons with great force in support of the same, without giving any solution, and therefore have been judged grievously suspected of heresy; that is to say, that I held and believed that the sun is the centre of the world and immoveable, and that the earth is not the centre and moveable, Willing, therefore, to remove from the minds of Your Eminences, and of every Catholic Christian, this vehement suspicion rightfully entertained towards me, with a sincere heart and unfeigned faith, I abjure, curse, and detest, the said errors and heresies, and generally every other error and sect contrary to the said Holy Church; and I swear, that I will never more in future say or assert anything verbally, or in writing, which may give rise to a similar suspicion of me: but if I shall know any heretic, or any one suspected of heresy, that I will denounce him to this Holy Office, or to the Inquisitor and Ordinary of the place in which I may be. I swear, moreover, and promise, that I will fulfil, and observe fully, all the penances which have been, or shall be laid on me by this Holy Office. But if it shall happen that I violate any of my said, promises, oaths, and protestations, (which God avert!) I subject myself to all the pains and punishments, which have been decreed and promulgated by the sacred canons, and other general and particular constitutions, against delinquents of this description. So may God help me, and his Holy Gospels, which I touch with my own hands. I, the above-named Galileo Galilei, have abjured, sworn, promised, and bound myself, as above, and in witness thereof with my own hand have subscribed this present writing of my abjuration, which

---

* Page 18.

I have recited word for word. At Rome in the Convent of Minerva, 22d June, 1633. I, Galileo Galilei, have abjured as above with my own hand."
It is said that Galileo, as he rose from his knees, stamped on the ground, and whispered to one of his friends, *E pur si muove*—(It does move though).

Copies of Galileo's sentence and abjuration were immediately promulgated in every direction, and the professors at several universities received directions to read them publicly. At Florence this ceremony took place in the church of Sta. Croce, whither Guiducci, Aggiunti, and all others who were known in that city as firm adherents to Galileo's opinions, were specially summoned. The triumph of the "Paper Philosophers" was so far complete, and the alarm occasioned by this proof of their dying power extended even beyond Italy. "I have been told," writes Descartes from Holland to Mersenne at Paris, " that Galileo's system was printed in Italy last year, but that every copy has been burnt at Rome, and himself condemned to some sort of penance, which has astonished me so much that I have almost determined to burn all my papers, or at least never to let them be seen by any one. I cannot collect that he, who is an Italian and even a friend of the Pope, as I understand, has been criminated on any other account than for having attempted to establish the motion of the earth. I know that this opinion was formerly censured by some Cardinals, but I thought I had since heard, that no objection was now made to its being publicly taught, even at Rome."

The sentiments of all who felt themselves secured against the apprehension of personal danger could take but one direction, for, as Pascal well expressed it in one of his celebrated letters to the Jesuits—" It is in vain that you have procured against Galileo a decree from Rome condemning his opinion of the earth's motion. Assuredly, that will never prove it to be at rest; and if we have unerring observations proving that it turns round, not all mankind together can keep it from turning, nor themselves from turning with it."

The assembly of doctors of the Sorbonne at Paris narrowly escaped from passing a similar sentence upon the system of Copernicus. The question was laid before them by Richelieu, and it appears that their opinion was for a moment in favour of confirming the Roman decree. It is to be wished that the name had been preserved of one of its members, who, by his strong and philosophical representations, saved that celebrated body from this disgrace.

Those who saw nothing in the punishment of Galileo but passion and blinded superstition, took occasion to revert to the history of a similar blunder of the Court of Rome in the middle of the eighth century. A Bavarian bishop, named Virgil, eminent both as a man of letters and politician, had asserted the existence of Antipodes, which excited in the ignorant bigots of his time no less alarm than did the motion of the earth in the seventeenth century. Pope Zachary, who was scandalized at the idea of another earth, inhabited by another race of men, and enlightened by another sun and moon (for this was the shape which Virgil's system assumed in his eyes), sent out positive orders to his legate in Bavaria. "With regard to Virgil, the philosopher, (I know not whether to call him priest,) if he own these perverse opinions, strip him of his priesthood, and drive him from the church and altars of God." But Virgil had himself occasionally acted as legate, and was moreover too necessary to his sovereign to be easily displaced. He utterly disregarded these denunciations, and during twenty-five years which elapsed before his death, retained his opinions, his bishopric of Salzburg, and his political power. He was afterwards canonized*.

Even the most zealous advocates of the authority of Rome were embarrassed in endeavouring to justify the treatment which Galileo experienced. Tiraboschi has attempted to draw a somewhat subtle distinction between the bulls of the Pope and the inquisitorial decrees which were sanctioned and approved by him; he dwells on the reflection that no one, even among the most zealous Catholics, has ever claimed infallibility as an attribute of the Inquisition, and looks upon it as a special mark of grace accorded to the Roman Catholic Church, that during the whole period in which most theologians rejected the opinions of Copernicus, as contrary to the Scriptures, the head of that Church was never permitted to compromise his infallible character by formally condemning it †.

Whatever may be the value of this

---

* Annalium Bolorum, libri vii. Ingolstadii, 1554.
† La Chiesa non ha mai dichiarati eretici i sostenitori del Sistema Copernicano, e questa troppo rigorosa censura non uscì che dal tribunale della Romana Inquisizione a cui niuno tra Cattolici ancor piu zelanti ha mai attribuito il diritto dell' infalli-

consolation, it can hardly be conceded, unless it be at the same time admitted that many scrupulous members of the Church of Rome have been suffered to remain in singular misapprehension of the nature and sanction of the authority to which Galileo had yielded. The words of the bull of Sixtus V., by which the Congregation of the Index was remodelled in 1588, are quoted by a professor of the University of Louvain, a zealous antagonist of Galileo, as follows: "They are to examine and expose the books which are repugnant to the Catholic doctrines and Christian discipline, and after reporting on them to us, they are to condemn them by our authority.*" Nor does it appear that the learned editors of what is commonly called the Jesuit's edition of Newton's "Principia" were of opinion, that in adopting the Copernican system they should transgress a mandate emanating from any thing short of infallible wisdom. The remarkable words which they were compelled to prefix to their book, show how sensitive the court of Rome remained, even so late as 1742, with regard to this rashly condemned theory. In their preface they say: "Newton in this third book supposes the motion of the earth. We could not explain the author's propositions otherwise than by making the same supposition. We are therefore forced to sustain a character which is not our own; but we profess to pay the obsequious reverence which is due to the decrees pronounced by the supreme Pontiffs against the motion of the earth."†

This coy reluctance to admit what nobody any longer doubts has survived to the present time; for Bailli informs us,‡ that the utmost endeavours of Lalande, when at Rome, to obtain that Galileo's work should be erased from the Index, were entirely ineffectual, in consequence of the decree which had been fulminated against him; and in fact both it, and the book of Copernicus, "Nisi Corrigatur," are still to be seen on the forbidden list of 1828.

The condemnation of Galileo and his book was not thought sufficient. Urban's indignation also vented itself upon those who had been instrumental in obtaining the licence for him. The Inquisitor at Florence was reprimanded; Riccardi, the master of the sacred palace, and Ciampoli, Urban's secretary, were both dismissed from their situations. Their punishment appears rather anomalous and inconsistent with the proceedings against Galileo, in which it was assumed that his book was not properly licensed; yet the others suffered on account of granting that very licence; which he was accused of having surreptitiously obtained from them, by concealing circumstances with which they were not bound to be otherwise acquainted. Riccardi, in exculpation of his conduct, produced a letter in the hand-writing of Ciampoli, in which was contained that His Holiness, in whose presence the letter professed to be written, ordered the licence to be given. Urban only replied that this was a Ciampolism; that his secretary and Galileo had circumvented him; that he had already dismissed Ciampoli, and that Riccardi must prepare to follow him. As soon as the ceremony of abjuration was concluded, Galileo was consigned, pursuant to his sentence, to the prison of the Inquisition. Probably it was never intended that he should long remain there, for at the end of four days, he was reconducted on a very slight representation of Nicolini to the ambassador's palace, there to await his further destination. Florence was still suffering under the before-mentioned contagion; and Sienna was at last fixed on as the place of his relegation. He would have been shut up in some convent in that city, if Nicolini had not recommended as a more suitable residence, the palace of the Archishop Piccolomini, whom he knew to be among Galileo's warmest friends. Urban consented to the change, and Galileo finally left Rome for Sienna in the early part of July.

Piccolomini received him with the utmost kindness, controlled of course by the strict injunctions which were dispatched from Rome, not to suffer him on any account to quit the confines of the palace. Galileo continued at Sienna in this state of seclusion till December of the same year, when the contagion having ceased in Tuscany, he applied for permission to return to his villa at Arcetri. This was allowed, subject to the same restrictions under which he had been residing with the archbishop.

---

bilità. Anzi in cio ancora è d' ammirarsi la provi_denza di Dio à favor della Chiesa, percioche in un tempo in cui la maggior parte dei teologi ferma_mente credavano che il Sistema Copernicano fosse all' autorità delle sacre Carte contrario, pur non permise che dalla Chiesa si proferisse su cio un solenne giudizio.—Stor. della Lett. Ital.

* Lib. Fromondi Antaristarchus, Antwerpiæ, 1631.
† Newtoni Principia, Coloniæ, 1760.
‡ Histoire de l'Astronomie Moderne.

Chapter XIV.
Extracts from the Dialogues on the System.

After narrating the treatment to which Galileo was subject on account of his admirable Dialogues, it will not be irrelevant to endeavour, by a few extracts, to convey some idea of the style in which they are written. It has been mentioned, that he is considered to surpass all other Italian writers (unless we except Machiavelli) in the purity and beauty of his language, and indeed his principal followers, who avowedly imitated his style, make a distinguished group among the classical authors of modern Italy. He professed to have formed himself from the study of Ariosto, whose poems he passionately admired, insomuch that he could repeat the greater part of them, as well as those of Berni and Petrarca, all which he was in the frequent habit of quoting in conversation. The fashion and almost universal practice of that day was to write on philosophical subjects in Latin; and although Galileo wrote very passably in that language, yet he generally preferred the use of Italian, for which he gave his reasons in the following characteristic manner:—

"I wrote in Italian because I wished every one to be able to read what I wrote; and for the same cause I have written my last treatise in the same language: the reason which has induced me is; that I see young men brought together indiscriminately to study to become physicians, philosophers, &c., and whilst many apply to such professions who are most unfit for them, others who would be competent remain occupied either with domestic business, or with other employments alien to literature; who, although furnished, as Ruzzante might say, with a *decent set of brains;* yet, not being able to understand things written in *gibberish,* take it into their heads, that in these crabbed folios there must be some grand *hocus pocus* of logic and philosophy much too high up for them to think of jumping at. I want them to know, that as Nature has given eyes to them just as well as to philosophers for the purpose of seeing her works, she has also given them brains for examining and understanding them."

The general structure of the dialogues has been already described*; we shall therefore premise no more than the judgment pronounced on them by a highly gifted writer, to supply the deficiencies of our necessarily imperfect analysis.

"One forms a very imperfect idea of Galileo, from considering the discoveries and inventions, numerous and splendid as they are, of which he was the undisputed author. It is by following his reasonings, and by pursuing the train of his thoughts, in his own elegant, though somewhat diffuse exposition of them, that we become acquainted with the fertility of his genius—with the sagacity, penetration, and comprehensiveness of his mind. The service which he rendered to real knowledge is to be estimated, not only from the truths which he discovered, but from the errors which he detected—not merely from the sound principles which he established, but from the pernicious idols which he overthrew. The dialogues on the system are written with such singular felicity, that one reads them at the present day, when the truths contained in them are known and admitted, with all the delight of novelty, and feels one's self carried back to the period when the telescope was first directed to the heavens, and when the earth's motion, with all its train of consequences, was proved for the first time."*

The first Dialogue is opened by an attack upon the arguments by which Aristotle pretended to determine *à priori* the necessary motions belonging to different parts of the world, and on his favourite principle that particular motions belong naturally to particular substances. Salviati (representing Galileo) then objects to the Aristotelian distinctions between the corruptible elements and incorruptible skies, instancing among other things the solar spots and newly appearing stars, as arguments that the other heavenly bodies may probably be subjected to changes similar to those which are continually occurring on the earth, and that it is the great distance alone which prevents their being observed. After a long discussion on this point, Sagredo exclaims, "I see into the heart of Simplicio, and perceive that he is much moved by the force of these too conclusive arguments; but methinks I hear him say—'Oh, to whom must we betake ourselves to settle our disputes if Aristotle be removed from the chair? What

---

* See page 56.

* Playfair's Dissertation, Supp. Encyc. Brit.

F

other author have we to follow in our schools, our studies, and academies? What philosopher has written on all the parts of Natural Philosophy, and so methodically as not to have overlooked a single conclusion? Must we then desolate this fabric, by which so many travellers have been sheltered? Must we destroy this asylum, this Prytaneum wherein so many students have found a convenient resting-place, where without being exposed to the injuries of the weather, one may acquire an intimate knowledge of nature, merely by turning over a few leaves? Shall we level this bulwark, behind which we are safe from every hostile attack? I pity him no less than I do one who at great expense of time and treasure, and with the labour of hundreds, has built up a very noble palace; and then, because of insecure foundations, sees it ready to fall—unable to bear that those walls be stripped that are adorned with so many beautiful pictures, or to suffer those columns to fall that uphold the stately galleries, or to see ruined the gilded roofs, the chimney-pieces, the friezes, and marble cornices erected at so much cost, he goes about it with girders and props, with shores and buttresses, to hinder its destruction."

Salviati proceeds to point out the many points of similarity between the earth and moon, and among others which we have already mentioned, the following remark deserves especial notice:—

"Just as from the mutual and universal tendency of the parts of the earth to form a whole, it follows that they all meet together with equal inclination, and that they may unite as closely as possible, assume the spherical form; why ought we not to believe that the moon, the sun, and other mundane bodies are also of a round figure, from no other reason than from a common instinct and natural concourse of all their component parts; of which if by accident any one should be violently separated from its whole, is it not reasonable to believe that spontaneously, and of its natural instinct, it would return? It may be added that if any centre of the universe may be assigned, to which the whole terrene globe if thence removed would seek to return, we shall find most probable that the sun is placed in it, as by the sequel you shall understand."

Many who are but superficially acquainted with the History of Astronomy, are apt to suppose that Newton's great merit was in his being the first to suppose an attractive force existing in and between the different bodies composing the solar system. This idea is very erroneous; Newton's discovery consisted in conceiving, and proving the identity of the force with which a stone falls, and that by which the moon falls, towards the earth (on an assumption that this force becomes weaker in a certain proportion as the distance increases at which it operates), and in generalizing this idea, in applying it to all the visible creation, and tracing the principle of universal gravitation with the assistance of a most refined and beautiful geometry into many of its most remote consequences. But the general notion of an attractive force between the sun, moon, and planets, was very commonly entertained before Newton was born, and may be traced back to Kepler, who was probably the first modern philosopher who suggested it. The following extraordinary passages from his "Astronomy" will shew the nature of his conceptions on this subject:—

"The true doctrine of gravity is founded on these axioms: every corporeal substance, so far forth as it is corporeal, has a natural fitness for resting in every place where it may be situated by itself beyond the sphere of influence of its cognate body. Gravity is a mutual affection between cognate bodies towards union or conjunction (similar in kind to the magnetic virtue), so that the earth attracts a stone much rather than the stone seeks the earth. Heavy bodies (if in the first place we put the earth in the centre of the world) are not carried to the centre of the world in its quality of centre of the world, but as to the centre of a cognate round body, namely the earth. So that wheresoever the earth may be placed or whithersoever it may be carried by its animal faculty, heavy bodies will always be carried towards it. If the earth were not round heavy bodies would not tend from every side in a straight line towards the centre of the earth, but to different points from different sides. If two stones were placed in any part of the world near each other and beyond the sphere of influence of a third cognate body, these stones, like two magnetic needles, would come together in the intermediate point, each approaching the other by a space proportional to the comparative mass of the other. If the moon and earth were not, retained in their orbits by their animal force or some other equivalent, the earth would mount to the moon by a fifty-fourth part of their distance, and the moon fall towards the earth through the other fifty-three parts, and would there meet, assuming however that the substance of both is of the same density. If the earth should cease to attract its waters to itself, all the waters of the sea would be raised, and would flow to the body of the moon*."

He also conjectured that the irregularities in the moon's motion were caused by the joint action of the sun and earth, and recognized the mutual action of the sun and planets, when he declared the mass and density of the sun to be so great that the united attraction of the other planets cannot remove it from its place.. Among these bold and brilliant ideas, his temperament led him to introduce others which show how unsafe it was to follow his guidance, and which account for, if they do not altogether justify, the sarcastic remark of Ross, that "Kepler's opinion that the planets are moved round by the sunne, and that this is done by sending forth a magnetic virtue, and that the sun-beames are like the teethe of a wheele taking hold of the planets, are senslesse crotchets fitter for a wheeler or a miller than a philosopher."† Roberval took up Kepler's notions, especially in the tract which he falsely attributed to Aristarchus, and it is much to be regretted that Roberval should deserve credit for anything connected with that impudent fraud. The principle of universal gravitation, though not the varying proportion, is distinctly assumed in it, as the following passages will sufficiently prove: " In every single particle of the earth, and the terrestrial elements, is a certain property or accident which we suppose common to the whole system of the world, by virtue of which all its parts are forced together, and reciprocally attract each other ; and this property is found in a greater or less degree in the different particles, according to their density. If the earth be considered by itself, its centres of magnitude and virtue, or gravity, as we usually call it, will coincide, to which all its parts tend in a straight line, as well, by their own exertion or gravity, as by the reciprocal attraction of all the rest." In a subsequent chapter, Roberval repeats these passages nearly in the same words, applying them to the whole solar system, adding, that " the force of this attraction, is not to be considered as residing in the centre itself, as some ignorant people think, but in the whole system whose parts are equally disposed round the centre*". This very curious work was reprinted in the third volume of the *Reflexiones Physico-Mathematicæ* of Mersenne, from whom Roberval pretended to have received the Arabic manuscript, and who is thus irretrievably implicated in the forgery.† The last remark, denying the attractive force to be due to any property of the central point, seems aimed at Aristotle, who, in a no less curious passage, maintaining exactly the opposite opinion, says, " Hence, we may better understand what the ancients have related, that like things are wont to have a tendency, to each other. For this is not absolutely true ; for if the earth were to be removed to the place now occupied by the moon, no part of the earth would then have a tendency towards that place, but would still fall towards the point which the earth's centre now occupies."‡ Mersenne considered the consequences of the attractive force of each particle of matter so far as to remark, that if a body were supposed to fall towards the centre of the earth, it would be retarded by the attraction of the part through which it had already fallen.§ Galileo had not altogether neglected to speculate on such a supposition, as is plain from the following extract. It is taken from a letter to Carcaville, dated from Arcetri, in 1637. " I will say farther, that I have not absolutely and clearly satisfied myself that a heavy body would arrive sooner at the centre of the earth, if it began to fall from the distance only of a single yard, than another which should start from the distance of a thousand miles. I do not affirm this, but I offer it as a paradox." ¶

It is very difficult to offer any satisfactory comment upon this passage ; it may be sufficient to observe that this paradoxical result was afterwards de-

---

* Astronomia NoVa. Pragæ. 1609.
† The new Planet no Planet, or the Earth no wandering Star, except in the wandering heads of Galileans. London, 1646.

* Aristarchi Samii de Mundi Systemate. Parisiis 1644.
† See page 12.
‡ De Cœlo, lib. iv. cap. 3.
§ Reflexiones Physico-Mathematicæ. Parisiis, 1647.
¶ Venturi.

duced by Newton, as one of the cousequences of the general law with which all nature is pervaded, but with which there is no reason to believe that Galileo had any acquaintance; indeed the idea is fully negatived by other passages in this same letter. This is one of the many instances from which we may learn to be cautious how we invest detached passages of the earlier mathematicians with a meaning which in many cases their authors did not contemplate. The progressive development of these ideas in the hands of Wallis, Huyghens, Hook, Wren, and Newton, would lead us too far from our principal subject. There is another passage in the third dialogue connected with this subject, which it may be as well to notice in this place. "The parts of the earth have such a propensity to its centre, that when it changes its place, although they may be very distant from the globe at the time of the change, yet must they follow. An example similar to this is the perpetual sequence of the Medicean stars, although always separated from Jupiter. The same may be said of the moon, obliged to follow the earth. And this may serve for those simple ones who have difficulty in comprehending how these two globes, not being chained together, nor strung upon a pole, mutually follow each other, so that on the acceleration or retardation of the one, the other also moves quicker or slower."

The second Dialogue is appropriated chiefly to the discussion of the diurnal motion of the earth; and the principal arguments urged by Aristotle, Ptolemy, and others, are successively brought forward and confuted. The opposers of the earth's diurnal motion maintained, that if it were turning round, a stone dropped from the top of a tower would not fall at its foot; but, by the rotation of the earth to the eastward carrying away the tower with it, would be left at a great distance to the westward; it was common to compare this effect to a stone dropped from the mast-head of a ship, and without any regard to truth it was boldly asserted that this would fall considerably nearer the stern than the foot of the mast, if the ship were in rapid motion. The same argument was presented in a variety of forms,—such as that a cannon-ball shot perpendicularly upwards would not fall at the same spot; that if fired to the eastward it would fly farther than to the westward; that a mark to the east or west would never be hit, because of the rising or sinking of the horizon during the flight of the ball; that ladies ringlets would all stand out to the westward,* with other conceits of the like nature: to which the general reply is given, that in all these cases the stone, or ball, or other body, participates equally in the motion of the earth, which, therefore, so far as regards the relative motion of its parts, may be disregarded. The manner in which this is illustrated, appears in the following extract from the dialogue :—*Sagredo.* If the nib of a writing pen which was in the ship during my voyage direct from Venice to Alexandria, had had the power of leaving a visible mark of all its path, what trace, what mark, what line would it have left?—"*Simplicio.* It would have left a line stretched out thither from Venice not perfectly straight, or to speak more correctly, not perfectly extended in an exact circular arc, but here and there more and less curved accordingly as the vessel had pitched more or less; but this variation in some places of one or two yards to the right or left, or up or down in a length of many hundred miles, would have occasioned but slight alteration in the whole course of the line, so that it would have been hardly sensible, and without any great error we may speak of it as a perfectly circular arc.— *Sagred.* So that the true and most exact motion of the point of the pen, would also have been a perfect arc of a circle if the motion of the vessel, abstracting from the fluctuations of the waves, had been steady and gentle; and if I had held this pen constantly in my hand, and had merely moved it an inch or two one way or the other, what alteration would that have made in the true and principal motion ?—*Simpl.* Less than that which would be occasioned in a line a thousand yards long, by varying here and there from perfect straightness by the quantity of a flea's eye.—*Sagred.* If then a painter on our quitting the port had begun to draw with this pen on paper, and had continued his drawing till we got to Alexandria, he would have been able by its motion, to produce an accurate representation of many objects perfectly shadowed, and filled up on all sides with landscapes, buildings, and animals, although all the true, real, and essential motion of the point of his pen would have been no other but a very

---

* Riccioli.

long and very simple line; and as to the peculiar work of the painter, he would have drawn it exactly the same if the ship had stood still. Therefore, of the very protracted motion of the pen, there remain no other traces than those marks drawn upon the paper, the reason of this being that the great motion from Venice to Alexandria was common to the paper, the pen, and everything that was in the ship; but the trifling motion forwards and backwards, to the right and left, communicated by the painter's fingers to the pen, and not to the paper, from being peculiar to the pen, left its mark upon the paper, which as to this motion was immoveable. Thus it is likewise true that in the supposition of the earth's rotation, the motion of a falling stone is really a long track of many hundreds and thousands of yards; and if it could have delineated its course in the calm air, or on any other surface, it would have left behind it a very long transversal line; but that part of all this motion which is common to the stone, the tower, and ourselves, is imperceptible by us and the same as if not existing, and only that part remains to be observed of which neither we nor the tower partake, which in short is the fall of the stone along the tower."

The mechanical doctrines introduced into this second dialogue will be noticed on another occasion; we shall pass on to other extracts, illustrative of the general character of Galileo's reasoning:—
"*Salviati.* I did not say that the earth has no principle, either internal or external, of its motion of rotation, but I do say that I know not which of the two it has, and that my ignorance has no power to take its motion away; but if this author knows by what principle other mundane bodies, of the motion of which we are certain, are turned round, I say that what moves the Earth is something like that by which Mars and Jupiter, and, as he believes, the starry sphere, are moved round; and if he will satisfy me as to the cause of their motion, I bind myself to be able to tell him what moves the earth. Nay more; I undertake to do the same if he can teach me what it is which moves the parts of the earth downwards.— *Simpl.* The cause of this effect is notorious, and every one knows that it is Gravity.—*Salv.* You are out, Master Simplicio; you should say that every one knows that it is called Gravity; but I do not ask you the name but the nature of the thing, of which nature you do not know one tittle more than you know of the nature of the moving cause of the rotation of the stars, except it be the name which has been given to the one, and made familiar and domestic, by the frequent experience we have of it many thousand times in a day; but of the principle or virtue by which a stone falls to the ground, we really know no more than we know of the principle which carries it upwards when thrown into the air, or which carries the moon round its orbit, except, as I have said, the name of gravity which we have peculiarly and exclusively assigned to it; whereas we speak of the other with a more generic term, and talk of the virtue impressed, and call it either an assisting or an informing intelligence, and are content to say that Nature is the cause of an infinite number of other motions."

Simplicio is made to quote a passage from Scheiner's book of Conclusions against Copernicus, to the following effect:—" 'If the whole earth and water were annihilated, no hail or rain would fall from the clouds, but would only be naturally carried round in a circle, nor would any fire or fiery thing ascend, since, according to the not improbable opinion of these others, there is no fire in the upper regions.'—*Salv.* The foresight of this philosopher is most admirable and praiseworthy, for he is not content with providing for things that might happen during the common course of nature, but persists in shewing his care for the consequences of what he very well knows will never come to pass. Nevertheless, for the sake of hearing some of his notable conceits, I will grant that if the earth and water were annihilated there would be no more hail or rain, nor would fiery matter ascend any more, but would continue a motion of revolution. What is to follow? What conclusion is the philosopher going to draw?—*Simpl.* This objection is in the very next words— 'Which nevertheless (says he) is contrary to experience and reason.'—*Salv.* Now I must yield: since he has so great an advantage over me as experience, with which I am quite unprovided. For hitherto I have never happened to see the terrestial earth and water annihilated, so as to be able to observe what the hail and fire did in the confusion. But does he tell us for our information at least what they did?—*Simp.* No, he does not say any thing more.—

Salv. I would give something to have a word or two with this person, to ask him whether, when this globe vanished, it also carried away the common centre of gravity, as I fancy it did, in which case I take it that the hail and water would remain stupid and confounded amongst the clouds, without knowing what to do with themselves. . . . And lastly, that I may give this philosopher a less equivocal answer, I tell him that I know as much of what would follow after the annihilation of the terrestrial globe, as he could have known what was about to happen in and about it, before it was created."

Great part of the third Dialogue is taken up with discussions on the parallax of the new stars of 1572 and 1604, in which Delambre notices that Galileo does not employ logarithms in his calculations, although their use had been known since Napier discovered them in 1616: the dialogue then turns to the annual motion " first taken from the Sun and conferred upon the Earth' by Aristarchus Samius, and afterwards by Copernicus." Salviati speaks of his contemporary philosophers with great contempt—" If you had ever been worn out as I have been many and many a time with hearing what sort of stuff is sufficient to make the obstinate vulgar unpersuadable, I do not say to agree with, but even to listen to these novelties, I believe your wonder at finding so few followers of these opinions would greatly fall off. But little regard in my judgment is to be had of those understandings who are convinced and immoveably persuaded of the fixedness of the earth, by seeing that they are not able to breakfast this morning at Constantinople, and sup in the evening in Japan, and who feel satisfied that the earth, so heavy as it is, cannot climb up above the sun, and then come tumbling in a breakneck fashion down again!"* This remark serves to introduce several specious arguments against the annual motion of the earth, which are successively confuted, and it is shewn how readily the apparent stations and retrogradations of the planets are accounted for on this supposition.

The following is one of the frequently recurring passages in which Galileo, whilst arguing in favour of the enormous distances at which the theory of Copernicus necessarily placed the fixed stars, inveighs against the arrogance with which men pretend to judge of matters removed above their comprehension. " *Simpl.* All this is very well, and it is not to be denied that the heavens may surpass in bigness the capacity of our imaginations, as also that God might have created it yet a thousand times larger than it really is, but we ought not to admit anything to be created in vain, and useless in the universe. Now whilst we see this beautiful arrangement of the planets, disposed round the earth at distances proportioned to the effects they are to produce on us for our benefit, to what purpose should a vast vacancy be afterwards interposed between the orbit of Saturn and the starry spheres, containing not a single star, and altogether useless and unprofitable ? to what end? for whose use and advantage ?—*Salv.* Methinks we arrogate too much to ourselves, Simplicio, when we will have it that the care of us alone is the adequate and sufficient work and bound, beyond which the divine wisdom and power does and disposes of nothing. I feel confident that nothing is omitted by the Divine Providence of what concerns the government of human affairs; but that there may not be other things in the universe dependant upon His supreme wisdom, I cannot for myself, by what my reason holds out to me, bring myself to believe. So that when I am told of the uselessness of an immense space interposed between the orbits of the planets and the fixed stars, empty and valueless, I reply that there is temerity in attempting by feeble reason to judge the works of God, and in calling vain and superfluous every part of the universe which is of no use to us.—*Sagr.* Say rather, and I believe you would say better, that we have no means of knowing what is of use to us; and I hold it to be one of the greatest pieces of arrogance and folly that can be in this world to say, because I know not of what use Jupiter or Saturn are to me, that therefore these planets are superfluous; nay more, that there are no such things in nature. To understand what effect is worked upon us by this or that heavenly body (since you will have it that all their use must have a reference to us), it would be necessary to remove it for a

---

* The notions commonly entertained of ' up' and ' down,' as connected with the observer's own situation, had long been a stumbling-block in the way of the new doctrines. When Columbus held out the certainty of arriving in India by sailing to the westward on account of the earth's roundness, it was gravely objected, that it might be well enough to sail down to India, but that the chief difficulty would consist in climbing up back again.

while, and then the effect which I find no longer produced in me, I may say that it depended upon that star. Besides, who will dare say that the space which they call too vast and useless between Saturn and the fixed stars is void of other bodies belonging to the universe. Must it be so because we do not see them: then I suppose the four Medicean planets, and the companions of Saturn, came into the heavens when we first began to see them, and not before! and, by the same rule, the other innumerable fixed stars did not exist before men saw them. The nebulæ were till lately only white flakes, till with the telescope we have made of them constellations of bright and beautiful stars. Oh presumptuous! rather, Oh rash ignorance of man!"

After a discussion on Gilbert's Theory of Terrestrial Magnetism, introduced by the parallelism of the earth's axis, and of which Galileo praises very highly both the method and results, the dialogue proceeds as follows:—"*Simpl.* It appears to me that Sig. Salviati, with a fine circumlocution, has so clearly explained the cause of these effects, that any common understanding, even though unacquainted with science, may comprehend it: but we, confining ourselves to the terms of art, reduce the cause of these and other similar natural phenomena to sympathy, which is a certain agreement and mutual appetency arising between things which have the same qualities, just as, on the other hand, that disagreement and aversion, with which other things naturally repel and abhor each other, we style antipathy.—*Sagr.* And thus with these two words they are able to give a reason for the great number of effects and accidents which we see, not without admiration, to be produced in Nature. But it strikes me that this mode of philosophising has a great sympathy with the style in which one of my friends used to paint: on one part of the canvas he would write with chalk—there I will have a fountain,with Diana and her nymphs.; here some harriers ; in this corner I will have a huntsman, with a stag's head; the rest may be a landscape of wood and mountain ; and what remains to be done may be put in by the colourman: and thus he flattered himself that he had painted the story of Actæon, having contributed nothing to it beyond the names."

The fourth Dialogue is devoted entirely to an examination of the tides, and is a development and extension of the treatise already mentioned to have been sent to the Archduke Leopold, in 1618*. Galileo was uncommonly partial to his theory of the tides, from which he thought to derive a direct proof of the earth's motion in her orbit ; and although his theory was erroneous, it required a farther advance in the science of motion than had been attained even at a much later period to point out the insufficiency of it. It is well known that the problem of explaining the cause of this alternate motion of the waters had been considered from the earliest ages one of the most difficult that could be proposed, and the solutions with which different inquirers were obliged to rest contented, shew that it long deserved the name given to it, of "the grave of human curiosity†." Riccioli has enumerated several of the opinions which in turn had their favourers and supporters. One party supposed the rise of the waters to be occasioned by the influx of rivers into the sea ; others compared the earth to a large animal, of which the tides indicated the respiration ; a third theory supposed the existence of subterraneous fires, by which the sea was periodically made to boil ; others attributed the cause of a similar change of temperature to the sun and moon.

There is an unfounded legend, that Aristotle drowned himself in despair of being able to invent a plausible explanation of the extraordinary tides in the Euripus. His curiosity on the subject does not appear to have been so acute (judging from his writings) as this story would imply. In one of his books he merely mentions a rumour, that there are great elevations or swellings of the seas, which recur periodically, according to the course of the moon. Lalande, in the fourth volume of his Astronomy, has given an interesting account of the opinion of the connection of the tides with the moon's motion. Pytheas of Marseilles, a contemporary of Aristotle, was the first who has been recorded as observing, that the full tides occur at full moon, and the ebbs at new moon‡. This is not quite correctly stated; for the tide of new moon is known to be still higher than the rise at the full, but it is likely enough, that the seeming inaccuracy should be attributed, not to

---

* See page 50. † Riccioli Almag. Nov.
‡ Plutarch, De placit. Philos. lib. iii. c. 17.

Pytheas, but to his biographer Plutarch, who, in many instances, appears to have viewed the opinions of the old philosophers through the mist of his own prejudices and imperfect information. The fact is, that, on the same day when the tide rises highest, it also ebbs lowest; and Pytheas, who, according to Pliny, had recorded a tide in Britain of eighty cubits, could not have been ignorant of this. Posidonius, as quoted by Strabo, maintained the existence of three periods of the tide, daily, monthly, and annual, " in sympathy with the moon." * Pliny, in his vast collection of natural observations, not unaptly styled the Encyclopædia of the Antients, has the following curious passages:—
" The flow and ebb of the tide is very wonderful; it happens in a variety of ways, but the cause is in the sun and moon†." He then very accurately describes the course of the tide during a revolution of the moon, and adds: " The flow takes place every day at a different hour; being waited on by the star, which rises every day in a different place from that of the day before, and with greedy draught drags the seas with it‡." " When the moon is in the north, and further removed from the earth, the tides are more gentle than when digressing to the south, she exerts her force with a closer effort§."

The College of Jesuits at Coimbra appears to deserve the credit of first clearly pointing out the true relation between the tides and the moon, which was also maintained a few years later by Antonio de Dominis and Kepler. In the Society's commentary on Aristotle's book on Meteors, after refuting the notion that the tides are caused by the light of the sun and moon, they say, " It appears more probable to us, without any rarefaction, of which there appears no need or indication, that the moon raises the waters by some inherent power of impulsion, in the same manner as a magnet moves iron; and according to its different aspects and approaches to the sea, and the obtuse or acute angles of its bearing, at one time to attract and raise the waters along the shore, and then again to leave them to sink down by their own weight, and

to gather into a lower level.*" The theory of Universal Gravitation seems here within the grasp of these philosophers, but unfortunately it did not occur to them that possibly the same attraction might be exerted on the earth as well as the water, and that the tide was merely an effect of the diminution of force, owing to the increase of distance, with which the centre of the earth is attracted, as compared with that exerted on its surface. This idea, so happily seized afterwards by Newton, might at once have furnished them with a satisfactory explanation of the tide, which is observed on the opposite side of the earth as well as immediately under the moon. They might have seen that in the latter case the centre of the earth is pulled away from the water, just as in the former the water is pulled away from the centre of the earth, the sensible effect to us being in both cases precisely the same. For want of this generalization, the inferior tide as it is called presented a formidable obstacle to this theory, and the most plausible explanation that was given was, that this magnetic virtue radiated out from the moon was reflected by the solid heavens, and concentrated again as in a focus on the opposite side of the earth. The majority of modern astronomers who did not admit the existence of any solid matter fit for producing the effect assigned to it, found a reasonable difficulty in acquiescing in this explanation. Galileo, who mentions the Archbishop of Spalatro's book, treated the theory of attraction by the moon as absurd. "This motion of the seas is local and sensible, made in an immense mass of water, and cannot be brought to obey light, and warmth, and predominancy of occult qualities, and such like vain fancies; all which are so far from being the cause of the tide, that on the contrary the tide is the cause of them, inasmuch as it gives rise to these ideas in brains which are more apt for talkativeness and ostentation, than for speculation and inquiry into the secrets of Nature; who, rather than see themselves driven to pronounce these wise, ingenuous, and modest words—*I do not know*,—will blurt out from their tongues and pens all sorts of extravagancies."

Galileo's own theory is introduced by the following illustration, which indeed

---

* συμπαθέως τῃ σελήνῃ. Geographiæ, lib. iii.
† Historia Naturalis, lib. ii. c. 97.
‡ Ut ancillante sidere, trahenteque secum avido haustu maria.
§ Eādem Aquilonia, et à terris longius recedente, mitiores quam cum, in Austros digressā, propiore nisu vim suam exercet.

* Commentarii Collegii Conimbricensis. Coloniæ, 1603.

probably suggested it, as he was in the habit of suffering no natural phenomena, however trivial in appearance, to escape him. He felt the advantage of this custom in being furnished on all occasions with a stock of homely illustrations, to which the daily experience of his hearers readily assented, and which he could shew to be identical in principle with the phenomena under discussion. That he was mistaken in applying his observations in the present instance cannot be urged against the incalculable value of such a habit.

"We may explain and render sensible these effects by the example of one of those barks which come continually from Lizza Fusina, with fresh water for the use of the city of Venice. Let us suppose one of these barks to come thence with moderate velocity along the canal, carrying gently the water with which it is filled, and then, either by touching the bottom, or from some other hindrance which is opposed to it, let it be notably retarded; the water will not on that account lose like the bark the impetus it has already acquired, but will forthwith run on towards the prow where it will sensibly rise, and be depressed at the stern. If on the contrary the said vessel in the middle of its steady course shall receive a new and sensible increase of velocity, the contained water before giving into it will persevere for some time in its slowness, and will be left behind that is to say towards the stern where consequently it will rise, and sink at the head.—Now, my masters, that which the vessel does in respect of the water contained in it, and that which the water does in respect of the vessel containing it, is the same to a hair as what the Mediterranean vase does in respect of the water which it contains, and that the waters do in respect of the Mediterranean vase which contains them. We have now only to demonstrate how, and in what manner it is true that the Mediterranean, and all other gulfs, and in short all the parts of the earth move with a motion sensibly not uniform, although no motion results thence to the whole globe which is not perfectly uniform and regular."

This unequable motion is derived from a combination of the earth's motion on her axis, and in her orbit, the consequence of which is that a point under the sun is carried in the same direction by the annual and diurnal velocities, whereas a point on the opposite side of the globe is carried in opposite directions by the annual and diurnal motions, so that in every twenty-four hours the absolute motion through space of every point in the earth completes a cycle of varying swiftness. Those readers who are unacquainted with the mathematical theory of motion must be satisfied with the assurance that this specious representation is fallacious, and that the oscillation of the water does not in the least result from the causes here assigned to it: the reasoning necessary to prove this is not elementary enough to be introduced here with propriety.

Besides the principal daily oscillation of the water, there is a monthly inequality in the rise and fall, of which the extremes are called the spring and neap tides: the manner in which Galileo attempted to bring his theory to bear upon these phenomena is exceedingly curious.

"It is a natural and necessary truth, that if a body be made to revolve, the time of revolution will be greater in a greater circle than in a less: this is universally allowed, and fully confirmed by experiments, such for instance as these:—In wheel clocks, especially in large ones, to regulate the going, the workmen fit up a bar capable of revolving horizontally, and fasten two leaden weights to the ends of it; and if the clock goes too slow, by merely approaching these weights somewhat towards the centre of the bar, they make its vibrations more frequent, at which time they are moving in smaller circles than before*.—Or, if you fasten a weight to a cord which you pass round a pulley in the ceiling, and whilst the weight is vibrating draw in the cord towards you, the vibrations will become sensibly accelerated as the length of the string diminishes. We may observe the same rule to hold among the celestial motions of the planets, of which we have a ready instance in the Medicean planets, which revolve in such short periods round Jupiter. We may therefore safely conclude, that if the moon for instance shall continue to be forced round by the same moving power, and were to move in a smaller circle, it would shorten the time of its revolution. Now this very thing happens in fact to the moon, which I have just advanced on a supposition. Let us call

---

* See fig. 1. p. 96.

to mind that we have already concluded with Copernicus, that it is impossible to separate the moon from the earth, round which without doubt it moves in a month: we must also remember that the globe of the earth, accompanied always by the moon, revolves in the great circle round the sun in a year, in which time the moon revolves round the earth about thirteen times, whence it follows that the moon is sometimes near the sun, that is to say between the earth and sun, sometimes far from it, when she is on the outside of the earth. Now if it be true that the power which moves the earth and the moon round the sun remains of the same efficacy, and if it be true that the same moveable, acted on by the same force, passes over similar arcs of circles in a time which is least when the circle is smallest, we are forced to the conclusion that at new moon, when in conjunction with the sun, the moon passes over greater arcs of the orbit round the sun, than when in opposition at full moon; and this inequality of the moon will be shared by the earth also. So that exactly the same thing happens as in the balance of the clocks; for the moon here represents the leaden weight, which at one time is fixed at a greater distance from the centre to make the vibrations slower, and at another time nearer to accelerate them."

Wallis adopted and improved this theory in a paper which he inserted in the Philosophical Transactions for 1666, in which he declares, that the circular motion round the sun should be considered as taking place at a point which is the centre of gravity of the earth and moon. "To the first objection, that it appears not how two bodies that have no tie can have one common centre of gravity, I shall only answer, that it is harder to show how they have it, than that they have it*." As Wallis was perfectly competent from the time at which he lived, and his knowledge of the farthest advances of science in his time, to appreciate the value of Galileo's writings, we shall conclude this chapter with the judgment that he has passed upon them in the same paper. "Since Galileo, and after him Torricelli and others have applied mechanical principles to the solving of philosophical difficulties, natural philosophy is well known to have been rendered more intelligible, and to have made a much greater progress in less than a hundred years than before for many ages."

Chapter XV.

Galileo at Arcetri—Becomes Blind— Moon's Libration — Publication of the Dialogues on Motion.

We have already alluded to the imperfect state of the knowledge possessed with regard to Galileo's domestic life and personal habits; there is reason however to think that unpublished materials exist from which these outlines might be in part filled up. Venturi informs us that he had seen in the collection from which he derived a great part of the substance of his Memoirs of Galileo, about one hundred and twenty manuscript letters, dated between the years 1623 and 1633, addressed to him by his daughter Maria, who with her sister had attached herself to the convent of St. Matthew, close to Galileo's usual place of residence. It is difficult not to think that much interesting information might be obtained from these, with respect to Galileo's domestic character. The very few published extracts confirm our favourable impressions of it, and convey a pleasing idea of this his favourite daughter. Even when, in her affectionate eagerness to soothe her father's wounded feelings at the close of his imprisonment in Rome, she dwells with delight upon her hopes of being allowed to relieve him, by taking on herself the penitential recitations which formed a part of his sentence, the prevalent feeling excited in every one by the perusal must surely be sympathy with the filial tenderness which it is impossible to misunderstand.

The joy she had anticipated in again meeting her parent, and in compensating to him by her attentive affection the insults of his malignant enemies, was destined to be but of short duration. Almost in the same month in which Galileo returned to Arcetri she was seized with a fatal illness; and already in the beginning of April, 1634, we learn her death from the fruitless condolence of his friends. He was deeply and bitterly affected by this additional blow, which came upon him when he was himself in a weak and declining state of health, and his answers breathe a spirit of the most hopeless and gloomy despondency.

In a letter written in April to Boc-

---

* Phil. Trans., No. 16, August 1666.

chineri, his son's father-in-law, he says: "The hernia has returned worse than at first: my pulse is intermitting, accompanied with a palpitation of the heart; an immeasurable sadness and melancholy; an entire loss of appetite; I am hateful to myself; and in short I feel that I am called incessantly by my dear daughter. In this state, I do not think it advisable that Vincenzo should set out on his journey, and leave me, when every hour something may occur, which would make it expedient that he should be here." In this extremity of ill health, Galileo requested leave to go to Florence for the advantage of medical assistance; but far from obtaining permission, it was intimated that any additional importunities would be noticed by depriving him of the partial liberty he was then allowed to enjoy. After several years confinement at Arcetri, during the whole of which time he suffered from continual indisposition, the inquisitor Fariano wrote to him in 1638, that the Pope permitted his removal to Florence, for the purpose of recovering his health; requiring him at the same time to present himself at the Office of the Inquisition, where he would learn the conditions on which this favour had been granted. These were that he should neither quit his house nor receive his friends there; and so closely was the letter of these instructions adhered to, that he was obliged to obtain a special permission to go out to attend mass during Passion week. The strictness with which all personal intercourse with his friends was interrupted, is manifest by the result of the following letter from the Duke of Tuscany's secretary of state to Nicolini, his ambassador at Rome. "Signor Galileo Galilei, from his great age and the illnesses which afflict him, is in a condition soon to go to another world;· and although in this the eternal memory of his fame and value is already secured, yet his Highness is greatly desirous that the world should sustain as little loss as possible by his death; that his labours may not perish, but for the public good may be brought to that perfection which he will not be able to give them. He has in his thoughts many things worthy of him, which he cannot be prevailed on to communicate to any but Father Benedetto Castelli, in whom he has entire confidence. His Highness wishes therefore that you should see Castelli, and induce him to procure leave to come to Florence for a few months for this purpose, which his Highness has very much at heart; and if he obtains permission, as his Highness hopes, you will furnish him with money and every thing else he may require for his journey." Castelli, it will be remembered, was at this time salaried by the court of Rome. Nicolini answered that Castelli had been himself to the Pope to ask leave to go to Florence. Urban immediately intimated his suspicions that his design was to see Galileo, and upon Castelli's stating that certainly it would be impossible for him to refrain from attempting to see him, he received permission to visit him in the company of an officer of the Inquisition. At the end of some months Galileo was remanded to Arcetri, which he never again quitted.

In addition to his other infirmities, a disorder which some years before had affected the sight of his right eye returned in 1636; in the course of the ensuing year the other eye began to fail also, and in a few months he became totally blind. It would be difficult to find any even among those who are the most careless to make a proper use of the invaluable blessing of sight, who could bear unmoved to be deprived of it, but on Galileo the loss fell with peculiar and terrible severity; on him who had boasted that he would never cease from using the senses which God had given him, in declaring the glory of his works, and the business of whose life had been the splendid fulfilment of that undertaking. "The noblest eye is darkened," said Castelli, "which nature ever made: an eye so privileged, and gifted with such rare qualities, that it may with truth be said to have seen more than all of those who are gone, and to have opened the eyes of all who are to come." His own patience and resignation under this fatal calamity are truly wonderful; and·if occasionally a word of complaint escaped him, it was in the chastened tone of the following expressions—"Alas! your dear friend and servant Galileo has become totally and irreparably blind; so that this heaven, this earth, this universe, which with wonderful observations I had enlarged a hundred and thousand times beyond the belief of by-gone ages, henceforward for me is shrunk into the narrow space which I myself fill in it.—So it pleases God: it shall therefore please me also." Hopes were at first entertained by Galileo's friends, that the blindness was occasioned by cataracts, and that he might look forward to relief from the operation of couching; but it very soon appeared that the disorder was not in the humours of the eye, but in a cloudiness of the cornea, the symptoms of which all external remedies failed to alleviate.

As long as the power was left him, he had indefatigably continued his astronomical observations. Just before his sight began to decay, he had observed a new phenomenon in the moon, which is now known by the name of the moon's libration, the nature of which we will shortly explain. A remarkable circumstance connected with the moon's motion is, that the same side is always visible from the earth, showing that the moon turns once on her own axis in exactly the time of her monthly revolution.* But Galileo, who was by this time familiar with the whole of the moon's visible surface, observed that the above-mentioned effect does not accurately take place, but that a small part on either side comes alternately forward into sight, and then again recedes, according to the moon's various positions in the heavens. He was not long in deteeting one of the causes of this apparent libratory or rocking motion. It is partly occasioned by our distance as spectators from the centre of the earth, which is also the centre of the moon's motion. In consequence of this, as the moon rises in the sky we get an additional view of the lower half, and lose sight of a small part of the upper half which was visible to us while we were looking down upon her when low in the horizon. The other cause is not quite so simple, nor is it so certain that Galileo adverted to it: it is however readily intelligible even to those who are unacquainted with astronomy, if they will receive as a fact that the monthly motion of the moon is not uniform, but that she moves quicker at one time than another, whilst the motion of rotation on her own axis, like that of the earth, is perfectly uniform. A very little reflection will show that the observed phenomenon will necessarily follow. If the moon did not turn on her axis, every side of her would be successively presented, in the course of a month, towards the earth; it is the motion of rotation which tends to carry the newly discovered parts out of sight.

Let us suppose the moon to be in that part of her orbit where she moves with her average motion, and that she is moving towards the part where she moves most quickly. If the motion in the orbit were to remain the same all the way round, the motion of rotation would be just sufficient at every point to bring round the same part of the moon directly in front of the earth. But since, from the supposed point, the moon is moving for some time round the earth with a motion continually growing quicker, the motion of rotation is not sufficiently quick to carry out of sight the entire part discovered by the motion of translation. We therefore get a glimpse of a narrow strip on the side *from* which the moon is moving, which strip grows broader and broader, till she passes the point where she moves most swiftly, and reaches the point of average swiftness on the opposite side of her orbit. Her motion is now continually growing slower, and therefore from this point the motion of rotation is too swift, and carries too much out of sight, or in other words, brings into sight a strip on the side *towards* which the moon is moving. This increases till she passes the point of least swiftness, and arrives at the point from which we began to trace her course, and the phenomena are repeated in the same order.

This interesting observation closes the long list of Galileo's discoveries in the heavens. After his abjuration, he ostensibly withdrew himself in a great measure from his astronomical pursuits; and employed himself till 1636 principally with his Dialogues on Motion, the last work of consequence that he published. In that year he entered into correspondence with the Elzevirs, through his friend Micanzio, on the project of printing a complete edition of his writings. Among the letters which Micanzio wrote on the subject is one intimating that he had enjoyed the gratification, in his quality of Theologian to the Republic of Venice, of refusing his sanction to a work written against Galileo and Copernicus. The temper however in which this refusal was an-

---

* Frisi says that Galileo did not perceive this conclusion (Elogio del Galileo); but see The Dial. on the System, Dial. I. pp. 61, 62, 85. Edit. 1744. Plutarch says, (De Placitis Philos. lib. ii. c. 28,) that the Pythagoreans believed the moon to have inhabitants fifteen times as large as men, and that their day is fifteen times as long as ours. It seems probable, that the former of these opinions was engrafted on the latter, which is true, and implies a perception of the fact in the text.

nounced, contrasts singularly with that of the Roman Inquisitors. "A book was brought to me which a Veronese Capuchin has been writing, and wished to print, denying the motion of the earth. I was inclined to let it go, to make the world laugh, for the ignorant beast entitles every one of the twelve arguments which compose his book, ' An irrefragable and undeniable demonstration,' and then adduces nothing but such childish trash as every man of sense has long discarded. For instance, this poor animal understands so much geometry and mathematics, that he brings forward as a demonstration, that if the earth could move, having nothing to support it, it must necessarily fall. He ought to have added that then we should catch all the quails. But when I saw that he speaks indecently of you, and has had the impudence to put down an account of what passed lately, saying that he is in possession of the whole of your process and sentence, I desired the man who brought it to me to go and be hanged. But you know the ingenuity of impertinence; I suspect he will succeed elsewhere, because he is so enamoured of his absurdities, that he believes them more firmly than his Bible."

After Galileo's condemnation at Rome, he had been placed by the Inquisition in the list of authors the whole of whose writings, '*edita et edenda*,' were strictly forbidden. Micanzio could not even obtain permission to reprint the Essay on Floating Bodies, in spite of his protestations that it did not in any way relate to the Copernican theory. This was the greatest stigma with which the Inquisition were in the habit of branding obnoxious authors; and, in consequence of it, when Galileo had completed his Dialogues on Motion, he found great difficulty in contriving their publication, the nature of which may be learned from the account which Pieroni sent to Galileo of his endeavours to print them in Germany. He first took the manuscript to Vienna, but found that every book printed there must receive the approbation of the Jesuits; and Galileo's old antagonist, Scheiner, happening to be in that city, Pieroni feared lest he should interfere to prevent the publication altogether, if the knowledge of it should reach him. Through the intervention of Cardinal Dietrichstein, therefore got permission to have it printed at Olmutz, and that it should be approved by a Dominican, so as. to keep the whole business a secret from Scheiner and his party; but during this negociation the Cardinal suddenly died, and Pieroni being besides dissatisfied with the Olmutz type, carried back the manuscript to Vienna, from which he heard that Scheiner had gone into Silesia. A new approbation was there procured, and the work was just on the point of being sent to press, when the dreaded Scheiner re-appeared in Vienna, on which Pieroni again thought it advisable to suspend the impression till his departure. In the mean time his own duty as a military architect in the Emperor's service carried him to Prague, where Cardinal Harrach, on a former occasion, had offered him the use of the newly-erected University press. But Harrach happened not to be at Prague, and this plan like the rest became abortive. In the meantime Galileo, wearied with these delays, had engaged with Louis Elzevir, who undertook to print the Dialogues at Amsterdam.

It is abundantly evident from Galileo's correspondence that this edition was printed with his full concurrence, although, in order to obviate further annoyance, he pretended that it was pirated from a manuscript copy which he sent into France to the Comte de Noailles, to whom the work is dedicated. The same dissimulation had been previously thought necessary, on occasion of the Latin translation of "The Dialogues on the System," by Bernegger, which Galileo expressly requested through his friend Deodati, and of which he more than once privately signified his approbation, presenting the translator with a valuable telescope, although he publicly protested against its appearance. The story which Bernegger introduced in his preface, tending to exculpate Galileo from any share in the publication, is by his own confession a mere fiction. Noailles had been ambassador at Rome, and, by his conduct there, well deserved the compliment which Galileo paid him on the present occasion.

As an introduction to the account of this work, which Galileo considered the best he had ever produced, it will become necessary to premise a slight sketch of the nature of the mechanical philosophy which he found prevailing, nearly as it had been delivered by Aristotle, with the same view with which we introduced specimens of the astronomical opinions current when Galileo began to write on that subject: they serve to show the nature and objects of the reasoning which he had to oppose; and, without some exposition of them, the aim and value of many of his arguments would be imperfectly understood and appreciated.

Chapter XVI.

State of the Science of Motion before Galileo.

It is generally difficult to trace any branch of human knowledge up to its origin, and more especially when, as in the case of mechanics, it is very closely connected with the immediate wants of mankind. Little has been told to us when we are informed that so soon as a man might wish to remove a heavy stone, " he would be led, by natural instinct, to slide under it the end of some long instrument, and that the same instinct would teach him either to raise the further end, or to press it downwards, so as to turn round upon some support placed as near to the stone as possible*."

Montucla's history would have lost nothing in value, if, omitting " this philosophical view of the birth of the art," he had contented himself with his previous remark, that there can be little doubt that men were familiar with the use of mechanical contrivances long before the idea occurred of enumerating or describing them, or even of examining very closely the nature and limits of the aid they are capable of affording. The most careless observer indeed could scarcely overlook that the weights heaved up with a lever, or rolled along a slope into their intended places, reached them more slowly than those which the workmen could lift directly in their hands; but it probably needed a much longer time to enable them to see the exact relation which, in these and all other machines, exists between the increase of the power to move, and the decreasing swiftness of the thing moved.

In the preface to Galileo's Treatise on Mechanical Science, published in 1592, he is at some pains to set in a clear light the real advantages belonging to the use of machines, " which (says he) I have thought it necessary to do, because, if I mistake not, I see almost all mechanics deceiving themselves in the belief that, by the help of a machine, they can raise a greater weight than they are able to lift by the exertion of the same force without it.—Now if we take any determinate weight, and any force, and any distance whatever, it is beyond doubt that we can move the weight to that distance by means of that force; because even although the force may be exceedingly small, if we divide the weight into a number of fragments, each of which is not too much for our force, and carry these pieces one by one, at length we shall have removed the whole weight; nor can we reasonably say at the end of our work, that this great weight has been moved and carried away by a force less than itself, unless we add that the force has passed several times over the space through which the whole weight has gone but once. From which it appears that the velocity of the force (understanding by velocity the space gone through in a given time) has been as many times greater than that of the weight, as the weight is greater than the force: nor can we on that account say that a great force is overcome by a small one, contrary to nature: then only might we say that nature is overcome when a small force moves a great weight as swiftly as itself, which we assert to be absolutely impossible with any machine either already or hereafter to be contrived. But since it may occasionally happen that we have but a small force, and want to move a great weight without dividing it into pieces, then we must have recourse to a machine by means of which we shall remove the given weight, with the given force, through the required space. But nevertheless the force as before will have to travel over that very same space as many times repeated as the weight surpasses its power, so that, at the end of our work, we shall find that we have derived no other benefit from our machine than that we have carried away the same weight altogether, which if divided into pieces we could have carried without the machine, by the same force, through the same space, in the same time. This is one of the advantages of a machine, because it often happens that we have a lack of force but abundance of time, and that we wish to move great weights all at once."

This compensation of force and time has been fancifully personified by saying that Nature cannot be cheated, and in scientific treatises on mechanics, is called the " principle of virtual velocities," consisting in the theorem that two weights will balance each other on any

---

* Histoire des Mathématiques, vol. i. p. 97.

machine, no matter how complicated or intricate the connecting contrivances may be, when one weight bears to the other the same proportion that the space through which the latter would be raised bears to that through which the former would sink, in the first instant of their motion, if the machine were stirred by a third force. The whole theory of machines consists merely in generalizing and following out this principle into its consequences; combined, when the machines are in a state of motion, with another principle equally elementary, but to which our present subject does not lead us to allude more particularly.

The credit of making known the principle of virtual velocities is universally given to Galileo; and so far deservedly, that he undoubtedly perceived the importance of it, and by introducing it everywhere into his writings succeeded in recommending it to others; so that five and twenty years after his death, Borelli, who had been one of Galileo's pupils, calls it "that mechanical principle with which everybody is so familiar*," and from that time to the present it has continued to be taught as an elementary truth in most systems of mechanics. But although Galileo had the merit in this, as in so many other cases, of familiarizing and reconciling the world to the reception of truth, there are remarkable traces before his time of the employment of this same principle, some of which have been strangely disregarded. Lagrange asserts† that the ancients were entirely ignorant of the principle of virtual velocities, although Galileo, to whom he refers it, distinctly mentions that he himself found it in the writings of Aristotle. Montucla quotes a passage from Aristotle's Physics, in which the law is stated generally, but adds that he did not perceive its immediate application to the lever, and other machines. The passage to which Galileo alludes is in Aristotle's Mechanics, where, in discussing the properties of the lever, he says expressly, "the same force will raise a greater weight, in proportion as the force is applied at a greater distance from the fulcrum, and the reason, as I have already said, is because it describes a greater circle; and a weight which is farther removed from the centre is made to move through a greater space."‡

It is true, that in the last mentioned treatise, Aristotle has given other reasons which belong to a very different kind of philosophy, and which may lead us to doubt whether he fully saw the force of the one we have just quoted. It appeared to him not wonderful that so many mechanical paradoxes (as he called them) should be connected with circular motion, since the circle itself seemed of so paradoxical a nature. "For, in the first place, it is made up of an immoveable centre, and a moveable radius, qualities which are contrary to each other. 2dly. Its circumference is both convex and concave. 3dly. The motion by which it is described is both forward and backward, for the describing radius comes back to the place from which it started. 4thly. The radius is *one*; but every point of it moves in describing the circle with a different degree of swiftness."

Perhaps Aristotle may have borrowed the idea of virtual velocities, contrasting so strongly with his other physical notions, from some older writer; possibly from Archytas, who, we are told, was the first to reduce the science of mechanics to methodical order; * and who by the testimony of his countrymen was gifted with extraordinary talents, although none of his works have come down to us. The other principles and maxims of Aristotle's mechanical philosophy, which we shall have occasion to cite, are scattered through his books on Mechanics, on the Heavens, and in his Physical Lectures, and will therefore follow rather unconnectedly, though we have endeavoured to arrange them with as much regularity as possible.

After defining a body to be that which is divisible in every direction, Aristotle proceeds to inquire how it happens that a body has only the three dimensions of length, breadth, and thickness; and seems to think he has given a reason in saying that, when we speak of two things, we do not say "all," but "both," and three is the first number of which we say "all." † When he comes to speak of motion, he says, "If motion is not understood, we cannot but remain ignorant of Nature. Motion appears to be of the nature of continuous quantities, and in continuous quantity infinity first makes its appearance; so as to furnish some with a definition who say that con-

---

* De vi Percussionis. Bononiæ, 1667.
† Mec. Analyt.   ‡ Mechanica.

* Diog. Laert. In vit. Archyt.
† De Cœlo, lib. i. c. 1.

tinuous quantity is that which is infinitely divisible.—Moreover, unless there be time, space, and a vacuum, it is impossible that there should be motion*."—
Few propositions of Aristotle's physical philosophy are more notorious than his assertion that nature abhors a vacuum, on which account this last passage is the more remarkable, as he certainly did not go so far as to deny the existence of motion, and therefore asserts here the necessity of that of which he afterwards attempts to show the absurdity.—" Motion is the energy of what exists in power so far forth as so existing. It is that act of a moveable which belongs to its power of moving." † After struggling through such passages as the preceding we come at last to a resting-place.—" It is difficult to understand what motion is."—When the same question was once proposed to another Greek philosopher, he walked away, saying, " I cannot tell you, but I will show you;" an answer intrinsically worth more than all the subtleties of Aristotle, who was not humble-minded enough to discover that he was tasking his genius beyond the limits marked out for human comprehension.

He labours in the same manner and with the same success to vary the idea of space. He begins the next book with declaring, that " those who say there is a vacuum assert the existence of space; for a vacuum is space, in which there is no substance;" and after a long and tedious reasoning concludes that, " not only what space is, but also whether there be such a thing, cannot but be doubted."‡ Of time he is content to say merely, that " it is clear that time is not motion, but that without motion there would be no time ;" § and there is perhaps little fault to be found with this remark, understanding motion in the general sense in which Aristotle here applies it, of every description of change.

Proceeding after these remarks on the nature of motion in general to the motion of bodies, we are told that " all local motion is either straight, circular, or compounded of these two; for these two are the only simple sorts of motion. Bodies are divided into simple and concrete; simple bodies are those which have naturally a principle of motion, as fire and earth, and their kinds. By simple motion is meant the motion of a simple body." * By these expressions Aristotle did not mean that a simple body cannot have what he calls a compound motion, but in that case he called the motion violent or unnatural; this division of motion into natural and violent runs through the whole of the mechanical philosophy founded upon his principles. " Circular motion is the only one which can be endless;"† the reason of which is given in another place: for " that cannot be doing, which cannot be done; and therefore it cannot be that a body should be moving towards a point (*i. e.* the end of an infinite straight line) whither no motion is sufficient to bring it." ‡ Bacon seems to have had these passages in view when he indulged in the reflections which we have quoted in page 14. " There are four kinds of motion of one thing by another: Drawing, Pushing, Carrying, Rolling. Of these, Carrying and Rolling may be referred to Drawing and Pushing.§—The prime mover and the thing moved are always in contact."

The principle of the composition of motions is stated very plainly : " when a moveable is urged in two directions with motions bearing any ratio to each other, it moves necessarily in a straight line, which is the diameter of the figure formed by drawing the two lines of direction in that ratio ;"‖ and adds, in a singularly curious passage, " but when it is urged for any time with two motions which have an indefinitely small ratio one to another, the motion cannot be straight, so that a body describes a curve, when it is urged by two motions bearing an indefinitely small ratio one to another, and lasting an indefinitely small time." ¶

---

* Phys. lib. i. c. 3.
† Lib. iii. c. 2. The Aristotelians distinguished between things as existing in act or energy (ενεργεια) and things in capacity or power (δυναμις). For the advantage of those who may think the distinction worth attending to, we give an illustration of Aristotle's meaning, from a very acute and learned commentator:—" It (motion) is something more than dead capacity; something less than perfect actuality ; capacity roused, and striving to quit its latent character; not the capable brass, nor yet the actual statue, but the capacity in energy; that is to say, the brass in fusion while it is becoming the statue and is not yet become."—" The bow moves not because it may be bent, nor because it is bent; but the motion lies between ; lies in an imperfect and obscure union of the two together; is the actuality (if I may so say) even of capacity itself: imperfect and obscure, because such is capacity to which it belongs."—Harris, Philosophical Arrangements.
‡ Lib. iv. c. 1.   § Lib. iv. c. 11.

* De Cœlo, lib. i. c. 2.   † Phys. lib. viii. c. 8.
‡ De Cœlo, lib. i. c. 6.   § Phys. lib. vii. c. 2.
‖ Mechanica.
¶ Εαν δε εν μηδενι λογῳ φερηται δυο φορας

He seemed on the point of discovering some of the real laws of motion, when he was led to ask—" Why are bodies in motion more easily moved than those which are at rest?—And why does the motion cease of things cast into the air? Is it that the force has ceased which sent them forth, or is there a struggle against the motion, or is it through the disposition to fall, does it become stronger than the projectile force, or is it foolish to entertain doubts on this question, when the body has quitted the principle of its motion?" A commentator at the close of the sixteenth century says on this passage: "They fall because every thing recurs to its nature; for if you throw a stone a thousand times into the air, it will never accustom itself to move upwards." Perhaps we shall now find it difficult not to smile at the idea we may form of this luckless experimentalist, teaching stones to fly; yet it may be useful to remember that it is only because we have already collected an opinion from the results of a vast number of observations in the daily experience of life, that our ridicule would not be altogether misplaced, and that we are totally unable to determine by any kind of reasoning, unaccompanied by experiment, whether a stone thrown into the air would fall again to the earth, or move for ever upwards, or in any other conceivable manner and direction.

The opinion which Aristotle held, that motion must be caused by something in contact with the body moved, led him to his famous theory that falling bodies are accelerated by the air through which they pass. We will show how it was attempted to explain this process when we come to speak of more modern authors. He classed natural bodies into heavy and light, remarking at the same time that it is clear that " there are some bodies possessing neither gravity nor levity*." By light bodies he understood those which have a natural tendency to move from the earth, observing that " that which is lighter is not always light†." He maintained that the heavenly bodies were altogether devoid of gravity; and we have already had occasion to mention his assertion, that a large body falls faster than a small one in proportion to its weight*. With this opinion may be classed another great mistake, in maintaining that the same bodies fall through different mediums, as air or water, with velocities reciprocally proportional to their densities. By a singular inversion of experimental science, Cardan, relying on this assertion, proposed in the sixteenth century to determine the densities of air and water by observing the different times taken by a stone in falling through them†. Galileo inquired afterwards why the experiment should not be made with a cork, which pertinent question put an end to the theory.

There are curious traces still preserved in the poem of Lucretius of a mechanical philosophy, of which the credit is in general given to Democritus, where many principles are inculcated strongly at variance with Aristotle's notions. We find absolute levity denied, and not only the assertion that in a vacuum all things would fall, but that they would fall with the same velocity; and the inequalities which we observe are attributed to the right cause, the impediment of the air, although the error remains of believing the velocity of bodies falling through the air to be proportional to their weight‡. Such specimens of this earlier philosophy

---

κατα μηδενα χρονον, αδυνατον ευθειαν ειναι την φοραν. Εαν γαρ τινα λογον ενεχθη εν χρονω τινι τουτον αναγκη τον χρονον ευθειαν ειναι φοραν δια τα προειρημενα, ωστε περιφερες γινεται δυο φερομενον φορας εν μηδενι λογω μηδενα χρονον,—i. e.
$$v = \frac{ds}{dt}$$
* De Cœlo, lib. i. c. 3, † Lib. iv. c. 2

* Phys., lib. iv. c. 8. † De Proport. Basileæ, 1570.
‡ " Nunc locus est, ut opinor, in his illud quoque rebus
Confirmare tibi, nullam rem posse suâ vi
Corpoream sursum ferri, sursumque meare.—
Nec quom subsiliunt ignes ad tecta domorum,
Et celeri flammâ degustant tigna trabeisque
Sponte suâ facere id sine vi subicente putandum est.
—Nonne vides etiam quantâ vi tigna trabeisque
Respuat humor aquæ? Nam quod magi' mersimus altum
Directâ et magnâ vi multi pressimus ægre:—
Tam cupide sursum revomit magis atque remittit
Plus ut parte foras emergant, exsiliantque:
—Nec tamen hæc, quantu'st in sedubitamus, opinor,
Quinvacuum per inane deorsum cuncta ferantur,
Sic igitur debent flammæ quoque posse per auras
Aeris expressæ sursum subsidere, quamquam
Pondera quantum in se est deorsum deducere pugnent.
—Quod si forte aliquis credit Graviora potesse
Corpora, quo citius rectum per Inane feruntur,
—Avius a verâ longe ratione recedit.
Nam per Aquas quæcunque cadunt atque Aera deorsum
Hæc pro ponderibus casus celerare necesse 'st
Propterea quia corpus Aquæ, naturaque tenuis
Aeris haud possunt æque rem quamque morari:
Sed citius cedunt Gravioribus exsuperata.
At contra nulli de nullâ parte, neque ullo
Tempore Inane potest Vacuum subsistere reii
Quin, sua quod natura petit, considere pergat:
Omniâ quâ propter debent per Inane quietum
Æque ponderibus non æquis concita ferri."
De Rerum Natura, lib. ii, v. 184—239.

may well indispose us towards Aristotle, who was as successful in the science of motion as he was in astronomy in suppressing the knowledge of a theory so much sounder than that which he imposed so long upon the credulity of his blinded admirers.

An agreeable contrast to Aristotle's mystical sayings and fruitless syllogisms is presented in Archimedes' book on Equilibrium, in which he demonstrates very satisfactorily, though with greater cumbrousness of apparatus than is now thought necessary, the principal properties of the lever. This and the Treatise on the Equilibrium of Floating Bodies are the only mechanical works which have reached us of this writer, who was by common consent one of the most accomplished mathematicians of antiquity. Ptolemy the astronomer wrote also a Treatise on Mechanics, now lost, which probably contained much that would be interesting in the history of mechanics; for Pappus says, in the Preface to the Eighth Book of his Mathematical Collections: "There is no occasion for me to explain what is meant by a heavy, and what by a light body, and why bodies are carried up and down, and in what sense these very words 'up' and 'down' are to be taken, and by what limits they are bounded; for all this is declared in Ptolemy's Mechanics."* This book of Ptolemy's appears to have been also known by Eutocius, a commentator of Archimedes, who lived about the end of the fifth century of our era; he intimates that the doctrines contained in it are grounded upon Aristotle's; if so, its loss is less to be lamented. Pappus's own book deserves attention for the enumeration which he makes of the mechanical powers, namely, the wheel and axle, the lever, pullies, the wedge and the screw. He gives the credit to Hero and Philo of having shown, in works which have not reached us, that the theory of all these machines is the same. In Pappus we also find the first attempt to discover the force necessary to support a given weight on an inclined plane. This in fact is involved in the theory of the screw; and the same vicious reasoning which Pappus employs on this occasion was probably found in those treatises which he quotes with so much approbation. Numerous as are the faults of his pretended demonstration, it was received undoubtingly for a long period.

The credit of first giving the true theory of equilibrium on the inclined plane is usually ascribed to Stevin, although, as we shall presently show, with very little reason. Stevin supposed a chain to be placed over two inclined planes, and to hang down in the manner represented in the figure. He then urged that the chain would be in equilibrium; for otherwise, it would incessantly continue in motion, if there were any cause why it should begin to move. This being conceded, he remarks further, that the parts AD and BD are also in equilibrium, being exactly similar to each other; and therefore if they are taken away, the remaining parts AC and BC will also be in equilibrium. The weights of these parts are proportional to the lengths AC and BC; and hence Stevin concluded that two weights would balance on two inclined planes, which are to each other as the lengths of the planes included between the same parallels to the horizon.* This conclusion is the correct one, and there is certainly great ingenuity in this contrivance to facilitate the demonstration; it must not however be mistaken for an à priori proof, as it sometimes seems to have been: we should remember that the experiments which led to the principle of virtual velocities are also necessary to show the absurdity of supposing a perpetual motion, which is made the foundation of this theorem. That principle had been applied directly to determine the same proportion in a work written long before, where it has remained singularly concealed from the notice of most who have written on this subject. The book bears the name of Jordanus, who lived at Namur in the thirteenth century; but Commandine, who refers to it in his Commentary on Pappus, considers it as the work of an earlier period. The author takes the principle of virtual velocities for the groundwork of his explanations, both of the lever and inclined plane; the latter will not occupy much space, and in an historical point of view is too curious to be omitted.

---

* Math. Coll. Pisani, 1662.

* Œuvres Mathématiques, Leyde, 1634.

"*Quæst.* 10.—If two weights descend by paths of different obliquities, and the proportion be the same of the weights and the inclinations taken in the same order, they will have the same descending force. By the inclinations, I do not mean the angles, but the paths up to the point in which both meet the same perpendicular.* Let, therefore, e be the weight upon $d\,c$, and h upon $d\,a$, and let e be to h as $d\,c$ to $d\,a$. I say these weights, in this situation, are equally effective. Take $d\,k$ equally inclined with $d\,c$, and upon it a weight equal to e, which call 6. If possible let e descend to l, so as to raise h to m, and

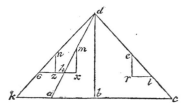

take 6 $\dot{n}$ equal to $h\,m$ or $e\,l$, and draw the horizontal and perpendicular lines as in the figure.
Then $n\,z : n\; 6 :: d\,b : d\,k$
and $m\,h : m\,x :: d\,a : d\,b$
therefore $n\,z : m\,x :: d\,a : d\,k :: h : 6$, *and therefore since* $e\,r$ *is not able to raise* 6 *to* n, *neither will it be able to raise* h *to* m; therefore they will remain as they are."† The passage in Italics tacitly assumes the principle in question. Tartalea, who edited Jordanus's book in 1565, has copied this theorem *verbatim* into one of his own treatises, and from that time it appears to have attracted no further attention. The rest of the book is of an inferior description. We find Aristotle's doctrine repeated, that the velocity of a falling body is proportional to its weight; that the weight of a heavy body changes with its form; and other similar opinions. The manner in which falling bodies are accelerated by the air is given in detail. "By its first motion the heavy body will drag after it what is behind, and move what is just below it; and these when put in motion move what is next to them, so that by being set in motion they less impede the falling body. In this manner it has the effect of being heavier, and impels still more those which give way before it, until at last they are no longer impelled, but begin to drag. And thus it happens that its gravity is increased by their attraction, and their motion by its gravity, whence we see that its velocity is continually multiplied."

In this short review of the state of mechanical science before Galileo, the name of Guido Ubaldi ought not to be omitted, although his works contain little or nothing original. We have already mentioned Benedetti as having successfully attacked some of Aristotle's statical doctrines, but it is to be noticed that the laws of motion were little if at all examined by any of these writers. There are a few theorems connected with this latter subject in Cardan's extraordinary book "On Proportions," but for the most part false and contradictory. In the seventy-first proposition of his fifth book, he examines the force of the screw in supporting a given weight, and determines it accurately on the principle of virtual velocities; namely, that the power applied at the end of the horizontal lever must make a complete circuit at that distance from the centre, whilst the weight rises through the perpendicular height of the thread. The very next proposition in the same page is to find the same relation between the power and weight on an inclined plane; and although the identity of principle in these two mechanical aids was well known, yet Cardan declares the necessary sustaining force to vary as the angle of inclination of the plane, for no better reason than that such an expression will properly represent it at the two limiting angles of inclination, since the force is nothing when the plane is horizontal, and equal to the weight when perpendicular. This again shows how cautious we should be in attributing the full knowledge of general principles to these early writers, on account of occasional indications of their having employed them.

CHAPTER XVII.
Galileo's theory of Motion—Extracts from the Dialogues.

DURING Galileo's residence at Sienna, when his recent persecution had rendered astronomy an ungrateful, and indeed an unsafe occupation for his ever active mind, he returned with increased pleasure to the favourite employment of

---

* This is not a literal translation, but by what follows, is evidently the Author's meaning. His words are, "Proportionem igitur declinationum dico non angulorum, sed linearum usque ad æquidistantem resecationem in quâ æqualiter sumunt de directo."

† Opusculum De Ponderositate, Venetiis, 1565.

his earlier years, an inquiry into the laws and phenomena of motion. His manuscript treatises on motion, written about 1590, which are mentioned by Venturi to be in the Ducal library at Florence, seem, from the published titles of the chapters, to consist principally of objections to the theory of Aristotle; a few only appear to enter on a new field of speculation. The 11th, 13th, and 17th chapters relate to the motion of bodies on variously inclined planes, and of projectiles. The title of the 14th implies a new theory of accelerated motion, and the assertion in that of the 16th, that a body falling naturally for however great a time would never acquire more than an assignable degree of velocity, shows that at this early period Galileo had formed just and accurate notions of the action of a resisting medium. It is hazardous to conjecture how much he might have then acquired of what we should now call more elementary knowledge; a safer course will be to trace his progress through existing documents in their chronological order. In 1602 we find Galileo apologizing in a letter addressed to his early patron the Marchese Guido Ubaldi, for pressing again upon his attention the isochronism of the pendulum, which Ubaldi had rejected as false and impossible. It may not be superfluous to observe that Galileo's results are not quite accurate, for there is a perceptible increase in the time occupied by the oscillations in larger arcs; it is therefore probable that he was induced to speak so confidently of their perfect equality, from attributing the increase of time which he could not avoid remarking to the increased resistance of the air during the larger vibrations. The analytical methods then known would not permit him to discover the curious fact, that the time of a total vibration is not sensibly altered by this cause, except so far as it diminishes the extent of the swing, and thus in fact, (paradoxical as it may sound) renders each oscillation successively more rapid, though in a very small degree. He does indeed make the same remark, that the resistance of the air will not affect the time of the oscillation, but that assertion was a consequence of his erroneous belief that the time of vibration in all arcs is the same. Had he been aware of the variation, there is no reason to think that he could have perceived that this result is not affected by it. In this letter is the first mention of the theorem, that the times of fall down all the chords drawn from the lowest point of a circle are equal; and another, from which Galileo afterwards deduced the curious result, that it takes less time to fall down the curve than down the chord, notwithstanding the latter is the direct and shortest course. In conclusion he says, "Up to this point I can go without exceeding the limits of mechanics, but I have not yet been able to demonstrate that all arcs are passed in the same time, which is what I am seeking." In 1604 he addressed the following letter to Sarpi, suggesting the false theory sometimes called Baliani's, who took it from Galileo.

"Returning to the subject of motion, in which I was entirely without a fixed principle, from which to deduce the phenomena I have observed, I have hit upon a proposition, which seems natural and likely enough; and if I take it for granted, I can show that the spaces passed in natural motion are in the double proportion of the times, and consequently that the spaces passed in equal times are as the odd numbers beginning from unity, and the rest. The principle is this, that the swiftness of the moveable increases in the proportion of its distance from the point whence it began to move; as for instance,—if a heavy body drop from A towards D, by the line A B C D, I suppose the degree of velocity which it has at B to bear to the velocity at C the ratio of A B to A C. I shall be very glad if your Reverence will consider this, and tell me your opinion of it. If we admit this principle, not only, as I have said, shall we demonstrate the other conclusions, but we have it in our power to show that a body falling naturally, and another projected upwards, pass through the same degrees of velocity. For if the projectile be cast up from D to A, it is clear that at D it has force enough to reach A, and no farther; and when it has reached C and B, it is equally clear that it is still joined to a degree of force capable of carrying it to A: thus it is manifest that the forces at D, C and B decrease in the proportion of A B, A C, and A D; so that if, in falling, the degrees of velocity observe the same proportion, that is true which I have hitherto maintained and believed."

We have no means of knowing how early Galileo discovered the fallacy of this reasoning. In his Dialogues on Motion, which contain the correct theory, he has put this erroneous supposition in the mouth of Sagredo, on which Salviati remarks, "Your discourse has so much likelihood in it, that our author himself did not deny to me when I proposed it to him, that he also had been for some time in the same mistake. But that which I afterwards extremely wondered at, was to see discovered in four plain words, not only the falsity, but the impossibility of a supposition carrying with it so much of seeming truth, that although I proposed it to many, I never met with any one but did freely admit it to be so; and yet it is as false and impossible as that motion is made in an instant: for if the velocities are as the spaces passed, those spaces will be passed in equal times, and consequently all motion must be instantaneous." The following manner of putting this reasoning will perhaps make the conclusion clearer. The velocity at any point is the space that would be passed in the next moment of time, if the motion be supposed to continue the same as at that point. At the beginning of the time, when the body is at rest, the motion is none; and therefore, on this theory, the space passed in the next moment is none, and thus it will be seen that the body cannot begin to move according to the supposed law.

A curious fact, noticed by Guido Grandi in his commentary on Galileo's Dialogues on Motion, is that this false law of acceleration is precisely that which would make a circular arc the shortest line of descent between two given points; and although in general Galileo only declared that the fall down the arc is made in less time than down the chord (in which he is quite correct), yet in some places he seems to assert that the circular arc is absolutely the shortest line of descent, which is not true. It has been thought possible that the law, which on reflection he perceived to be impossible, might have originally recommended itself to him from his perception that it satisfied his prejudice in this respect.

John Bernouilli, one of the first mathematicians in Europe at the beginning of the last century, has given us a proof that such a reason might impose even on a strong understanding, in the following argument urged by him in favour of Galileo's second and correct theory, that the spaces vary as the squares of the times. He had been investigating the curve of swiftest descent, and found it to be a cycloid, the same curve in which Huyghens had already proved that all oscillations are made in accurately equal times. "I think it," says he, "worthy of remark that this identity only occurs on Galileo's supposition, so that this alone might lead us to presume it to be the real law of nature. For nature, which always does everything in the very simplest manner, thus makes one line do double work, whereas on any other supposition, we must have had two lines, one for equal oscillations, the other for the shortest descent."*

Venturi mentions a letter addressed to Galileo in May 1609 by Luca Valerio, thanking him for his experiments on the descent of bodies on inclined planes. His method of making these experiments is detailed in the Dialogues on Motion:—" In a rule, or rather plank of wood, about twelve yards long, half a yard broad one way, and three inches the other, we made upon the narrow side or edge a groove of little more than an inch wide: we cut it very straight, and, to make it very smooth and sleek, we glued upon it a piece of vellum, polished and smoothed as exactly as possible, and in that we let fall a very hard, round, and smooth brass ball, raising one of the ends of the plank a yard or two at pleasure above the horizontal plane. We observed, in the manner that I shall tell you presently, the time which it spent in running down, and repeated the same observation again and again to assure ourselves of the time, in which we never found any difference, no, not so much as the tenth part of one heat of the pulse. Having made and settled this experiment, we let the same ball descend through a fourth part only of the length of the groove, and found the measured time to be exactly half the former. Continuing our experiments with other portions of the length, comparing the fall through the whole with the fall through half, two-thirds, three-fourths, in short, with the fall through any part, we found by many hundred experiments that the spaces passed over were as the squares of the times, and that this was the case in all inclinations of the plank; during which, we also re-

---

* Joh. Bernouilli, Opera Omnia, Lausannæ, 1744. tom. i. p. 192.

marked that the times of descent, on different inclinations, observe accurately the proportion assigned to them farther on, and demonstrated by our author. As to the estimation of the time, we hung up a great bucket full of water, which by a very small hole pierced in the bottom squirted out a fine thread of water, which we caught in a small glass during the whole time of the different descents: then weighing from time to time, in an exact pair of scales, the quantity of water caught in this way, the differences and proportions of their weights gave the differences and proportions of the times; and this with such exactness that, as I said before, although the experiments were repeated again and again, they never differed in any degree worth noticing." In order to get rid of the friction, Galileo afterwards substituted experiments with the pendulum; but with all his care he erred very widely in his determination of the space through which a body would fall in $1''$, if the resistance of the air and all other impediments were removed. He fixed it at 4 *braccia:* Mersenne has engraved the length of the ' *braccia* ' used by Galileo, in his " Harmonie Universelle," from which it appears to be about $23\frac{1}{2}$ English inches, so that Galileo's result is rather less than eight feet. Mersenne's own result from direct observation was thirteen feet: he also made experiments in St. Peter's at Rome, with a pendulum 325 feet long, the vibrations of which were made in $10''$; from this the fall in $1''$ might have been deduced rather more than sixteen feet, which is very close to the truth.

From another letter also written in the early part of 1609, we learn that Galileo was then busied with examining the strength and resistance " of beams of different sizes and forms, and how much weaker they are in the middle than at the ends, and how much greater weight they can support laid along their whole length, than if sustained on a single point, and of what form they should be so as to be equally strong throughout." He was also speculating on the motion of projectiles, and had satisfied himself that their motion in a vertical direction is unaffected by their horizontal velocity; a conclusion which, combined with his other experiments, led him afterwards to determine the path of a projectile in a non-resisting medium to be parabolical.

Tartalea is supposed to have been the first to remark that no bullet moves in a horizontal line; but his theory beyond this point was very erroneous, for he supposed the bullet's path through the air to be made up of an ascending and descending straight line, connected in the middle by a circular arc.

Thomas Digges, in his treatise on the Newe Science of Great Artillerie, came much nearer the truth; for he remarked[*], that " The bullet violentlye throwne out of the peece by the furie of the poulder hath two motions: the one violent, which endeuoreth to carry the bullet right out in his line diagonall, according to the direction of the peece's axis, from whence the violent motion proceedeth; the other naturall in the bullet itselfe, which endeuoreth still to carrye the same directlye downeward by a right line perpendiculare to the horizon, and which dooth though insensiblye euen from the beginning by little and little drawe it from that direct and diagonall course." And a little farther he observes that " These middle curve arkes of the bullet's circuite, compounded of the violent and naturall motions of the bullet, albeit they be indeed mere helicall, yet have they a very great resemblance of the Arkes Conical. And in randons above $45°$ they doe much resemble the Hyperbole, and in all vnder the Ellepsis. But exactlye they neuer accorde, being indeed Spirall mixte and Helicall."

Perhaps Digges deserves no greater credit from this latter passage than the praise of a sharp and accurate eye, for he does not appear to have founded this determination of the form of the curve on any theory of the direct fall of bodies; but Galileo's arrival at the same result was preceded, as we have seen, by a careful examination of the simplest phenomena into which this compound motion may be resolved. But it is time to proceed to the analysis of his " Dialogues on Motion," these preliminary remarks on their subject matter having been merely intended to show how long before their publication Galileo was in possession of the principal theories contained in them.

Descartes, in one of his letters to Mersenne, insinuates that Galileo had taken many things in these Dialogues from him: the two which he especially instances are the isochronism of the pendulum, and the law of the spaces varying

---

[*] Pantometria, 1591.

as the squares of the times.* Descartes was born in 1596: we have shown that Galileo observed the isochronism of the pendulum in 1583, and knew the law of the spaces in 1604, although he was then attempting to deduce it from an erroneous principle. As Descartes on more than one occasion has been made to usurp the credit due to Galileo, (in no instance more glaringly so than when he has been absurdly styled the forerunner of Newton,) it will not be misplaced to mention a few of his opinions on these subjects, recorded in his letters to Mersenne in the collection of his letters just cited:—" I am astonished at what you tell me of having found by experiment that bodies thrown up in the air take neither more nor less time to rise than to fall again; and you will excuse me if I say that I look upon the experiment as a very difficult one to make accurately. This proportion of increase according to the odd numbers 1, 3, 5, 7, &c., which is in Galileo, and which I think I wrote to you some time back, cannot be true, as I believe I intimated at the same time, unless we make two or three suppositions which are entirely false. One is Galileo's opinion, that motion increases gradually from the slowest degree; and the other is, that the air makes no resistance." In a later letter to the same person he says, apparently with some uneasiness, " I have been revising my notes on Galileo, in which I have not said expressly, that falling bodies do not pass through every degree of slowness, but I said that this cannot be determined without knowing what weight is; *which comes to the same thing.* As to your example, I grant that it proves that every degree of velocity is infinitely divisible, but not that a falling body actually passes through all these divisions.—It is certain that a stone is not equally disposed to receive a new motion or increase of velocity, when it is already moving very quickly, and when it is moving slowly. But I believe that I am now able to determine in what proportion the velocity of a stone increases, not when falling in a vacuum, but in this substantial atmosphere.— However I have now got my mind full of other things, and I cannot amuse myself with hunting this out, *nor is it a matter of much utility.*" He afterwards returns once more to the same subject:—" As to what Galileo says, that falling bodies pass through every degree of velocity, I do not believe that it generally happens, but I allow it is not impossible that it may happen occasionally." After this the reader will know what value to attach to the following assertion by the same Descartes:—" I see nothing in Galileo's books to envy him, and hardly any thing which I would own as mine;" and then may judge how far Salusbury's blunt declaration is borne out, " Where or when did any one appear that durst enter the lists with our Galileus? save only one bold and unfortunate Frenchman, who yet no sooner came within the ring but he was hissed out again."*

The principal merit of Descartes must undoubtedly be derived from the great advances he made in what are generally termed Abstract or Pure Mathematics; nor was he slow to point out to Mersenne and his other friends the acknowledged inferiority of Galileo to himself in this respect. We have not sufficient proof that this difference would have existed if Galileo's attention had been equally directed to that object; the singular elegance of some of his geometrical constructions indicates great talent for this as well as for his own more favourite speculations. But he was far more profitably employed: geometry and pure mathematics already far outstripped any useful application of their results to physical science, and it was the business of Galileo's life to bring up the latter to the same level. He found abstract theorems already demonstrated in sufficient number for his purpose, nor was there occasion to task his genius in search of new methods of inquiry, till all was exhausted which could be learned from those already in use. The result of his labours was that in the age immediately succeeding Galileo, the study of nature was no longer in arrear of the abstract theories of number and measure; and when the genius of Newton pressed it forward to a still higher degree of perfection, it became necessary to discover at the same time more powerful instruments of investigation. This alternating process has been successfully continued to the present time; the analyst acts as the pioneer of the naturalist, so that the abstract researches, which at first have no value but in the eyes of those to whom an elegant formula, in its own beauty, is a source of pleasure as real and as refined as a painting or a statue, are often found to furnish the

---

* Lettres de Descartes. Paris, 1657.

† Adam and Tannery vol. III, p. 9

* Math. Coll. vol. ii.

only means for penetrating into the most intricate and concealed phenomena of natural philosophy.

Descartes and Delambre agree in suspecting that Galileo preferred the dialogistic form for his treatises, because it afforded a ready opportunity for him to praise his own inventions: the reason which he himself gave is, the greater facility for introducing new matter and collateral inquiries, such as he seldom failed to add each time that he reperused his work. We shall select in the first place enough to show the extent of his knowledge on the principal subject, motion, and shall then allude as well as our limits will allow to the various other points incidentally brought forward.

The dialogues are between the same speakers as in the "System of the World;" and in the first Simplicio gives Aristotle's proof,* that motion in a vacuum is impossible, because according to him bodies move with velocities in the compound proportion of their weights and the rarities of the mediums through which they move. And since the density of a vacuum bears no assignable ratio to that of any medium in which motion has been observed, any body which should employ time in moving through the latter, would pass through the same distance in a vacuum instantaneously, which is impossible. Salviati replies by denying the axioms, and asserts that if a cannon ball weighing 200 lbs., and a musket ball weighing half a pound, be dropped together from a tower 200 yards high, the former will not anticipate the latter by so much as a foot; "and I would not have you do as some are wont, who fasten upon some saying of mine that may want a hair's breadth of the truth, and under this hair they seek to hide another man's blunder as big as a cable. Aristotle says that an iron ball weighing 100 lbs. will fall from the height of 100 yards while a weight of one pound falls but one yard: I say they will reach the ground together. They find the bigger to anticipate the less by two inches, and under these two inches they seek to hide Aristotle's 99 yards." In the course of his reply to this argument Salviati formally announces the principle which is the foundation of the whole of Galileo's theory of motion, and which must therefore be quoted in his own words:—"A heavy body has by nature an intrinsic principle of moving towards the common centre of heavy things; that is to say, to the centre of our terrestrial globe, with a motion continually accelerated in such manner that in equal times there are always equal additions of velocity. This is to be understood as holding true only when all accidental and external impediments are removed, amongst which is one that we cannot obviate, namely, the resistance of the medium. This opposes itself, less or more, accordingly as it is to open more slowly or hastily to make way for the moveable, which being by its own nature, as I have said, continually accelerated, consequently encounters a continually increasing resistance in the medium, until at last the velocity reaches that degree, and the resistance that power, that they balance each other; all further acceleration is prevented, and the moveable continues ever after with an uniform and equable motion." That such a limiting velocity is not greater than some which may be exhibited may be proved as Galileo suggested by firing a bullet upwards, which will in its descent strike the ground with less force than it would have done if immediately from the mouth of the gun; for he argued that the degree of velocity which the air's resistance is capable of diminishing must be greater than that which could ever be reached by a body falling naturally from rest. "I do not think the present occasion a fit one for examining the cause of this acceleration of natural motion, on which the opinions of philosophers are much divided; some referring it to the approach towards the centre, some to the continual diminution of that part of the medium remaining to be divided, some to a certain extrusion of the ambient medium, which uniting again behind the moveable presses and hurries it forwards. All these fancies, with others of the like sort, we might spend our time in examining, and with little to gain by resolving them. It is enough for our author at present that we understand his object to be the investigation and examination of some phenomena of a motion so accelerated, (no matter what may be the cause,) that the momenta of velocity, from the beginning to move from rest, increase in the simple proportion in which the time increases, which is as much as to say, that in equal times are equal additions of velocity. And if it shall turn out that the phenomena de-

---

* Phys. Lib. iv. c. 8.

monstrated on this supposition are verified in the motion of falling and naturally accelerated weights, we may thence conclude that the assumed definition does describe the motion of heavy bodies, and that it is true that their acceleration varies in the ratio of the time of motion."

When Galileo first published these Dialogues on Motion, he was obliged to rest his demonstrations upon another principle besides, namely, that the velocity acquired in falling down all inclined planes of the same perpendicular height is the same. As this result was derived directly from experiment, and from that only, his theory was so far imperfect till he could show its consistency with the above supposed law of acceleration. When Viviani was studying with Galileo, he expressed his dissatisfaction at this chasm in the reasoning; the consequence of which was, that Galileo, as he lay the same night, sleepless through indisposition, discovered the proof which he had long sought in vain, and introduced it into the subsequent editions. The third dialogue is principally taken up with theorems on the direct fall of bodies, their times of descent down differently inclined planes, which in planes of the same height he determined to be as the lengths, and with other inquiries connected with the same subject, such as the straight lines of shortest descent under different data, &c.

The fourth dialogue is appropriated to, projectile motion, determined upon the principle that the horizontal motion will continue the same as if there were no vertical motion, and the vertical motion as if there were no horizontal motion. "Let A B represent a horizontal

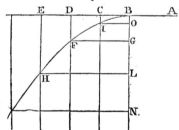

line or plane placed on high, on which let a body be carried with an equable motion from A towards B, and the support of the plane being taken away at B, let the natural motion downwards due to the body's weight come upon it in the direction of the perpendicular B N. Moreover let the straight line B E drawn in the direction A B be taken to represent the flow, or measure, of the time, on which let any number of equal parts B C, C D, D E, &c. be marked at pleasure, and from the points C, D, E, let lines be drawn parallel to B N; in the first of these let any part C I be taken, and let D F be taken four times as great as C I, E H nine times as great, and so on, proportionally to the squares of the lines B C, B D, B E, &c., or, as we say, in the double proportion of these lines. Now if we suppose that whilst by its equable horizontal motion the body moves from B to C, it also descends by its weight through C I, at the end of the time denoted by B C it will be at I. Moreover in the time B D, double of B C, it will have fallen four times as far, for in the first part of the Treatise it has been shewn that the spaces fallen through by a heavy body vary as the squares of the times. Similarly at the end of the time B E, or three times B C, it will have fallen through E H, and will be at H. And it is plain that the points I, F, H, are in the same parabolical line B I F H. The same demonstration will apply if we take any number of equal particles of time of whatever duration."

The curve called here a Parabola by Galileo, is one of those which results from cutting straight through a Cone, and therefore is called also one of the Conic Sections, the curious properties of which curves had drawn the attention of geometricians long before Galileo thus began to point out their intimate connexion with the phenomena of motion. After the proposition we have just extracted, he proceeds to anticipate some objections to the theory, and explains that the course of a projectile will not be accurately a parabola for two reasons; partly on account of the resistance of the air, and partly because a horizontal line, or one equi-distant from the earth's centre, is not straight, but circular. The latter cause of difference will, however, as he says, be insensible in all such experiments as we are able to make. The rest of the Dialogue is taken up with different constructions for determining the circumstances of the motion of projectiles, as their range, greatest height, &c.; and it is proved that, with a given force of projection, the range will be greatest when a ball is projected at an elevation of 45°, the ranges of all angles equally inclined above and below 45° corresponding exactly to each other.

One of the most interesting subjects discussed in these dialogues is the famous notion of Nature's horror of a vacuum or empty space, which the old school of philosophy considered as impossible to be obtained. Galileo's notions of it were very different; for although he still unadvisedly adhered to the old phrase to denote the resistance experienced in endeavouring to separate two smooth surfaces, he was so far from looking upon a vacuum as an impossibility, that he has described an apparatus by which he endeavoured to measure the force necessary to produce one. This consisted of a cylinder, into which is tightly fitted a piston; through the centre of the piston passes a rod with a conical valve, which, when drawn down, shuts the aperture closely, supporting a basket. The space between the piston and cylinder being filled full of water poured in through the aperture, the valve is closed, the vessel reversed, and weights are added till the piston is drawn forcibly downwards. Galileo concluded that the weight of the piston, rod, and added weights, would be the measure of the force of resistance to the vacuum which he supposed would take place between the piston and lower surface of the water. The defects in this apparatus for the purpose intended are of no consequence, so far as regards the present argument, and it is perhaps needless to observe that he was mistaken in supposing the water would not descend with the piston. This experiment occasions a remark from Sagredo, that he had observed that a lifting-pump would not work when the water in the cistern had sunk to the depth of thirty-five feet below the valve; that he thought the pump was injured, and sent for the maker of it, who assured him that no pump upon that construction would lift water from so great a depth. This story is sometimes told of Galileo, as if he had said sneeringly on this occasion that Nature's horror of a vacuum does not extend beyond thirty-five feet; but it is very plain that if he had made such an observation, it would have been seriously; and in fact by such a limitation he deprived the notion of the principal part of its absurdity. He evidently had adopted the common notion of suction, for he compares the column of water to a rod of metal suspended from its upper end, which may be lengthened till it breaks with its own weight. It is certainly very extraordinary that he failed to observe how simply these phenomena may be explained by a reference to the weight of the elastic atmosphere, which he was perfectly well acquainted with, and endeavoured by the following ingenious experiment to determine:—" Take a large glass flask with a bent neck, and round its mouth tie a leathern pipe with a valve in it, through which water may be forced into the flask with a syringe without suffering any air to escape, so that it will be compressed within the bottle. It will be found difficult to force in more than about three-fourths of what the flask will hold, which must be carefully weighed. The valve must then be opened, and just so much air will rush out as would in its natural density occupy the space now filled by the water. Weigh the vessel again; the difference will show the weight of that quantity of air*." By these means, which the modern experimentalist will see were scarcely capable of much accuracy, Galileo found that air was four hundred times lighter than water, instead of ten times, which was the proportion fixed on by Aristotle. The real proportion is about 830 times.

The true theory of the rise of water in a lifting-pump is commonly dated from Torricelli's famous experiment with a column of mercury, in 1644, when he found that the greatest height at which it would stand is fourteen times less than the height at which water will stand, which is exactly the proportion of weight between water and mercury. The following curious letter from Baliani, in 1630, shows that the original merit of suggesting the real cause belongs to him, and renders it still more unaccountable that Galileo, to whom it was addressed, should not at once have adopted the same view of the subject:— "I have believed that a vacuum may exist naturally ever since I knew that the air has sensible weight, and that you taught me in one of your letters how to find its weight exactly, though I have not yet succeeded with that experiment. From that moment I took up the notion

---

* It has been recently proposed to determine the density of high-pressure steam by a process analogous to this.

that it is not repugnant to the nature of things that there should be a vacuum, but merely that it is difficult to produce. To explain myself more clearly : if we allow that the air has weight, there is no difference between air and water except in degree. At the bottom of the sea the weight of the water above me compresses everything round my body, and it strikes me that the same thing must happen in the air, we being placed at the bottom of its immensity ; we do not feel its weight, nor the compression round us, because our bodies are made capable of supporting it. But if we were in a vacuum, then the weight of the air above our heads would be felt. It would be felt very great, but not infinite, and therefore determinable, and it might be overcome by a force proportioned to it. In fact I estimate it to be such that, to make a vacuum, I believe we require a force greater than that of a column of water thirty feet high*."

This subject is introduced by some observations on the force of cohesion, Galileo seeming to be of opinion that, although it cannot be adequately accounted for by " the great and principal resistance to a vacuum, yet that perhaps a sufficient cause may be found by considering every body as composed of very minute particles, between every two of which is exerted a similar resistance." This remark serves to lead to a discussion on indivisibles and infinite quantities, of which we shall merely extract what Galileo gives as a curious paradox suggested in the course of it. He supposes a basin to be formed by scooping a hemisphere out of a cylinder, and a cone to be taken of the same depth and base as the hemisphere. It is easy to show, if the cone and scooped cylinder be both supposed to be cut by the same plane, parallel to

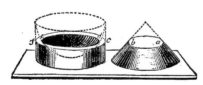

the one on which both stand, that the area of the ring C D E F thus discovered in the cylinder is equal to the area of the corresponding circular section AB of the cone wherever the cutting plane is supposed to be*. He then proceeds with these remarkable words:—" If we raise the plane higher and higher, one of these areas terminates in the circumference of a circle, and the other in a point, for such are the upper rim of the basin and the top of the cone. Now since in the diminution of the two areas they to the very last maintain their equality to one another, it is in my thoughts proper to say that the highest and ultimate terms † of such diminutions are equal, and not one infinitely bigger than the other. It seems therefore that the circumference of a large circle may be said to be equal to one single point. And why may not these be called equal if they be the last remainders and vestiges left by equal magnitudes ‡ ?"

We think no one can refuse to admit the probability, that Newton may have found in such passages as these the first germ of the idea of his prime and ultimate ratios, which afterwards became in his hands an instrument of such power. As to the paradoxical result, Descartes undoubtedly has given the true answer to it in saying that it only proves that the line is not a greater area than the point is. Whilst on this subject, it may not be uninteresting to remark that something similar to the doctrine of fluxions seems to have been lying dormant in the minds of the mathematicians of Galileo's era, for Inchoffer illustrates his argument in the treatise we have already mentioned, that the Copernicans may deduce some true results from what he terms their absurd hypothesis, by observing, that mathematicians may deduce the truth that a line is length without breadth, from the false and physically impossible supposition that a point flows, and that a line is the fluxion of a point §.

A suggestion that perhaps fire dissolves bodies by insinuating itself between their minute particles, brings on the subject of the violent effects of heat and light ; on which Sagredo inquires, whether we are to take for granted that the effect of light does or does not require time. Simplicio is ready with an answer, that the discharge of artillery proves the transmission of light to be

---

* Venturi, vol. ii.

* Galileo also reasons in the same way on the equality of the solids standing on the cutting plane, but one is sufficient for our present purpose.
† Gli altissimi e ultimi termini.
‡ Le ultime reliquie e vestigie lasciate da grandezze eguali.
§ Punctum fluere, et lineam esse fluxum puncti. Tract. Syllept. Romæ, 1633.

instantaneous, to which Sagredo cautiously replies, that nothing can be gathered from that experiment except that light travels more swiftly than sound; nor can we draw any decisive conclusion from the rising of the sun. "Who can assure us that he is not in the horizon before his rays reach our sight?" Salviati then mentions an experiment by which he endeavoured to examine this question. Two observers are each to be furnished with a lantern: as soon as the first shades his light, the second is to discover his, and this is to be repeated at a short distance till the observers are perfect in the practice. The same thing is to be tried at the distance of several miles, and if the first observer perceive any delay between shading his own light and the appearance of his companion's, it is to be attributed to the time taken by the light in traversing twice the distance between them. He allows that he could discover no perceptible interval at the distance of a mile, at which he had tried the experiment, but recommends that with the help of a telescope it should be tried at much greater distances. Sir Kenelm Digby remarks on this passage: "It may be objected (if there be some observable tardity in the motion of light) that the sunne would never be truly in that place in which unto our eyes he appeareth to be; because that it being seene by means of the light which issueth from it, if that light required time to move in, the sunne (whose motion is so swifte) would be removed from the place where the light left it, before it could be with us to give tidings of him. To this I answer, allowing peradventure that it may be so, who knoweth the contrary? Or what inconvenience would follow if it be admitted * ?"

The principal thing remaining to be noticed is the application of the theory of the pendulum to musical concords and dissonances, which are explained, in the same manner as by Kepler in his "Harmonices Mundi," to result from the concurrence or opposition of vibrations in the air striking upon the drum of the ear. It is suggested that these vibrations may be made manifest by rubbing the finger round a glass set in a large vessel of water; "and if by pressure the note is suddenly made to rise to the octave above, every one of the undulations which will be seen regularly spreading round the glass, will suddenly split into two, proving that the vibrations that occasion the octave are double those belonging to the simple note." Galileo then describes a method he discovered by accident of measuring the length of these waves more accurately than can be done in the agitated water. He was scraping a brass plate with an iron chisel, to take out some spots, and moving the tool rapidly upon the plate, he occasionally heard a hissing and whistling sound, very shrill and audible, and whenever this occurred, and then only, he observed the light dust on the plate to arrange itself in a long row of small parallel streaks equidistant from each other. In repeated experiments he produced different tones by scraping with greater or less velocity, and remarked that the streaks produced by the acute sounds stood closer together than those from the low notes. Among the sounds produced were two, which by comparison with a viol he ascertained to differ by an exact fifth; and measuring the spaces occupied by the streaks in both experiments, he found thirty of the one equal to forty-five of the other, which is exactly the known proportion of the lengths of strings of the same material which sound a fifth to each other *.

Salviati also remarks, that if the material be not the same, as for instance if it be required to sound an octave to a note on catgut, on a wire of the same length, the weight of the wire must be made four times as great, and so for other intervals. "The immediate cause of the forms of musical intervals is neither the length, the tension, nor the thickness, but the proportion of the numbers of the undulations of the air which strike upon the drum of the ear, and make it vibrate in the same intervals. Hence we may gather a plausible reason of the different sensations occasioned to us by different couples of sounds, of which we hear some with great pleasure, some with less, and call them accordingly concords, more or less perfect, whilst some excite in us great dissatisfaction, and are called discords. The disagreeable sensation belonging to the latter

---

* "Treatise of the Nature of Bodies. London, 1665."

* This beautiful experiment is more easily tried by drawing the bow of a violin across the edge of glass strewed with fine dry sand. Those who wish to see more on the subject may consult Chladni's 'Acoustique.'

probably arises from the disorderly manner in which the vibrations strike the drum of the ear; so that for instance a most cruel discord would be produced by sounding together two strings, of which the lengths are to each other as the side and diagonal of a square, which is the discord of the false fifth. On the contrary, agreeable consonances will result from those strings of which the numbers of vibrations made in the same time are commensurable, "to the end that the cartilage of the drum may not undergo the incessant torture of a double inflexion from the disagreeing percussions." Something similar may be exhibited to the eye by hanging up pendulums of different lengths: "if these be proportioned so that the times of their vibrations correspond with those of the musical concords, the eye will observe with pleasure their crossings and interweavings still recurring at appreciable intervals; but if the times of vibration be incommensurate, the eye will be wearied and worn out with following them."

The second dialogue is occupied entirely with an investigation of the strength of beams, a subject which does not appear to have been examined by any one before Galileo beyond Aristotle's remark, that long beams are weaker, because they are at once the weight, the lever, and the fulcrum; and it is in the development of this observation that the whole theory consists. The principle assumed by Galileo as the basis of his inquiries is, that the force of cohesion with which a beam resists a cross fracture in any section may all be considered as acting at the centre of gravity of the section, and that it breaks always at the lowest point: from this he deduced that the effect of the weight of a prismatic beam in overcoming the resistance of one end by which it is fastened to a wall, varies directly as the square of the length, and inversely as the side of the base. From this it immediately follows, that if for instance the bone of a large animal be three times as long as the corresponding one in a smaller beast, it must be nine times as thick to have the same strength, provided we suppose in both cases that the materials are of the same consistence. An elegant result which Galileo also deduced from this theory, is that the form of such a beam, to be equally strong in every part, should be that of a parabolical prism, the vertex of the parabola being the farthest removed from the wall. As an easy mode of describing the parabolic curve for this purpose, he recommends tracing the line in which a heavy flexible string hangs. This curve is not an accurate parabola: it is now called a catenary; but it is plain from the description of it in the fourth dialogue, that Galileo was perfectly aware that this construction is only approximately true. In the same place he makes the remark, which to many is so paradoxical, that no force, however great, exerted in a horizontal direction, can stretch a heavy thread, however slender, into an accurately straight line.

The fifth and sixth dialogues were left unfinished, and annexed to the former ones by Viviani after Galileo's death: the fragment of the fifth, which is on the subject of Euclid's Definition of Ratio, was at first intended to have formed a part of the third, and followed the first proposition on equable motion: the sixth was intended to have embodied Galileo's researches on the nature and laws of Percussion, on which he was employed at the time of his death. Considering these solely as fragments, we shall not here make any extracts from them.

Chapter XVIII.
Correspondence on Longitudes.—Pendulum Clock.

In the spring of 1636, having finished his Dialogues on Motion, Galileo resumed the plan of determining the longitude by means of Jupiter's satellites. Perhaps he suspected something of the private intrigue which thwarted his former expectations from the Spanish government, and this may have induced him on the present occasion to negotiate the matter without applying for Ferdinaud's assistance and recommendation. Accordingly he addressed himself to Lorenz Real, who had been Governor General of the Dutch possessions in India, freely and unconditionally offering the use of his theory to the States General of Holland. Not long before, his opinion had been requested by the commissioners appointed at Paris to examine and report on the practicability of another method proposed by Morin,* which consisted in observing the distance of the moon from a known star. Morin was a French philosopher, prin-

---

* One of the Commissioners was the father of Blaise Pascal.

cipally known as an astrologer and zealous Anti-Copernican; but his name deserves to be recorded as undoubtedly one of the first to recommend a method, which, under the name of a Lunar distance, is now in universal practice.

The monthly motion of the moon is so rapid, that her distance from a given star sensibly varies in a few minutes even to the unassisted eye; and with the aid of the telescope, we can of course appreciate the change more accurately. Morin proposed that the distances of the moon from a number of fixed stars lying near her path in the heavens should be beforehand calculated and registered for every day in the year, at a certain hour, in the place from which the longitudes were to be reckoned, as for instance at Paris. Just as in the case of the eclipses of Jupiter's satellites, the observer, when he saw that the moon had arrived at the registered distance, would know the hour at Paris: he might also make allowance for intermediate distances. Observing at the same instant the hour on board his ship, the difference between the two would show his position in regard of longitude. In using this method as it is now practised, several modifications are to be attended to, without which it would be wholly useless, in consequence of the refraction of the atmosphere, and the proximity of the moon to the earth. Owing to the latter cause, if two spectators should at the same instant of time, but in different places, measure the distance of the moon in the East, from a star still more to the eastward, it would appear greater to the more easterly spectator than to the other observer, who as seen from the star would be standing more directly behind the moon. The mode of allowing for these alterations is taught by trigonometry and astronomy.

The success of this method depends altogether upon the exact knowledge which we now have of the moon's course, and till that knowledge was perfected it would have been found altogether illusory. Such in fact was the judgment which Galileo pronounced upon it. "As to Morin's book on the method of finding the longitude by means of the moon's motion, I say freely that I conceive this idea to be as accurate in theory, as fallacious and impossible in practice. I am sure that neither you nor any one of the other four gentlemen can doubt the possibility of finding the difference of longitude between two meridians by means of the moon's motion, provided we are sure of the following requisites: First, an Ephemeris of the moon's motion exactly calculated for the first meridian from which the others are to be reckoned; secondly, exact instruments, and convenient to handle, in taking the distance between the moon and a fixed star; thirdly, great practical skill in the observer; fourthly, not less accuracy in the scientific calculations, and astronomical computations; fifthly, very perfect clocks to number the hours, or other means of knowing them exactly, &c. Supposing, I say, all these elements free from error, the longitude will be accurately found; but I reckon it more easy and likely to err in all of these together, than to be practically right in one alone. Morin ought to require his judges to assign, at their pleasure, eight or ten moments of different nights during four or six months to come, and pledge himself to predict and assign by his calculations the distances of the moon at those determined instants from some star which would then be near her. If it is found that the distances assigned by him agree with those which the quadrant or sextant* will actually show, the judges would be satisfied of his success, or rather of the truth of the matter, and nothing would remain but to show that his operations were such as could be performed by men of moderate skill, and also practicable at sea as well as on land. I incline much to think that an experiment of this kind would do much towards abating the opinion and conceit which Morin has of himself, which appears to me so lofty, that I should consider myself the eighth sage, if I knew the half of what Morin presumes to know."

It is probable that Galileo was biassed by a predilection for his own method, on which he had expended so much time and labour; but the objections which he raises against Morin's proposal in the foregoing letter are no other than those to which at that period it was undoubtedly open. With regard to his own, he had already, in 1612, given a rough prediction of the course of Jupiter's satellites, which had been found to agree tolerably well with subsequent observations; and since that

---

* These instruments were very inferior to those now in use under the same name. See "Treatise on Opt. Instrum."

time, amid all his other employments, he had almost unintermittingly during twenty-four years continued his observations, for the sake of bringing the tables of their motions to as high a state of perfection as possible. This was the point to which the inquiries of the States in their answer to Galileo's frank proposal were principally directed. They immediately appointed commissioners to communicate with him, and report the various points on which they required information. They also sent him a golden chain, and assured him that in the case of the design proving successful, he should have no cause to complain of their want of gratitude and generosity. The commissioners immediately commenced an active correspondence with him, in the course of which he entered into more minute details with regard to the methods by which he proposed to obviate the practical difficulties of the necessary observations.

It is worth noticing that the secretary to the Prince of Orange, who was mainly instrumental in forming this commission, was Constantine Huyghens, father of the celebrated mathematician of that name, of whom it has been said that he seemed destined to complete the discoveries of Galileo; and it is not a little remarkable, that Huyghens nowhere in his published works makes any allusion to this connexion between his father and Galileo, not even during the discussion that arose some years later on the subject of the pendulum clock, which must necessarily have forced it upon his recollection.

The Dutch commissioners had chosen one of their number to go into Italy for the purpose of communicating personally with Galileo, but he discouraged this scheme, from a fear of its giving umbrage at Rome. The correspondence being carried on at so great a distance necessarily experienced many tedious delays, till in the very midst of Galileo's labours to complete his tables, he was seized with the blindness which we have already mentioned. He then resolved to place all the papers containing his observations and calculations for this purpose in the hands of Renieri, a former pupil of his, and then professor of mathematics at Pisa, who undertook to finish and to forward them into Holland. Before this was done, a new delay was occasioned by the deaths which speedily followed each other of every one of the four commissioners; and for two or three years the correspondence with Holland was entirely interrupted. Constantine Huyghens, who was capable of appreciating the value of the scheme, succeeded after some trouble in renewing it, but only just before the death of Galileo himself, by which of course it was a second time broken off; and to complete the singular series of obstacles by which the trial of this method was impeded, just as Renieri, by order of the Duke of Tuscany, was about to publish the ephemeris and tables which Galileo had entrusted to him, and which the Duke told Viviani he had seen in his possession, he also was attacked with a mortal malady; and upon his death the manuscripts were nowhere to be found, nor has it since been discovered what became of them. Montucla has intimated his suspicions that Renieri himself destroyed them, from a consciousness that they were insufficient for the purpose to which it was intended to apply them; a bold conjecture, and one which ought to rest upon something more than mere surmise: for although it may be considered certain, that the practical value of these tables would be very inconsiderable in the present advanced state of knowledge, yet it is nearly as sure that they were unique at that time, and Renieri was aware of the value which Galileo himself had set upon them, and should not be lightly accused of betraying his trust in so gross a manner. In 1665, Borelli calculated the places of the satellites for every day in the ensuing year, which he professed to have deduced (by desire of the Grand Duke) from Galileo's tables;* but he does not say whether or not these tables were the same that had been in Renieri's possession.

We have delayed till this opportunity to examine how far the invention of the pendulum clock belongs to Galileo. It has been asserted that the isochronism of the pendulum had been noticed by Leonardo da Vinci, but the passage on which this assertion is founded (as translated from his manuscripts by Venturi) scarcely warrants this conclusion. "A rod which engages itself in the opposite teeth of a spur-wheel can act like the arm of the balance in clocks, that is to say, it will act alternately, first on one side of the wheel, then on the opposite

---

* Theoricæ Mediceorum Planetarum, Florentiæ, 1666.

one, without interruption." If Da Vinci had constructed a clock on this principle, and recognized the superiority of the pendulum over the old balance, he would surely have done more than merely mention it as affording an unintermitted motion "like the arm of the balance." The use of the balance is supposed to have been introduced at least as early as the fourteenth century. Venturi mentions the drawing and description of a clock in one of the manuscripts of the King's Library at Paris, dated about the middle of the fifteenth century, which as he says nearly resembles a modern watch. The balance is there called "The circle fastened to the stem of the pallets, and moved by the force with it.* In that singularly wild and extravagant book, entitled "A History of both Worlds," by Robert Flud, are given two drawings of the wheel-work of the clocks and watches in use before the application of the pendulum.¹ An inspection of them will show how little remained to be done when the isochronism of the pendulum was discovered. *Fig.* 1. represents "the

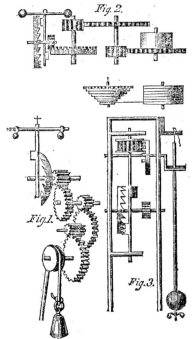

large clocks moved by a weight, such as are put up in churches and turrets;

* Circulus affixus virgæ paletorum qui cum eâ de vi movetur.

fig. 2. the small ones moved by a spring, such as are worn round the neck, or placed on a shelf or table. The use of the chain is to equalize the spring, which is strongest at the beginning of its motion."* This contrivance of the chain is mentioned by Cardan, in 1570, and is probably still older. In both figures the name given to the cross bar, with the weight attached to it, is "the time or balance (*tempus seu libratio*) by which the motion is equalized." The manner in which Huyghens first applied the pendulum is shown in *fig.* 3.† The action in the old clocks of the balance, or *rake*, as it was also called, was by checking the motion of the descending weight till its inertia was overcome; it was then forced round till the opposite pallet engaged in the toothed wheel. The balance was thus suddenly and forcibly reduced to a state of rest, and again set in motion in the opposite direction. It will be observed that these balances wanted the spiral spring introduced in all modern watches, which has a property of isochronism similar to that of the pendulum. Hooke is generally named as the discoverer of this property of springs, and as the author of its application to the improvement of watches, but the invention is disputed with him by Huyghens. Lahire asserts‡ that the isochronism of springs was communicated to Huyghens at Paris by Hautefeuille, and that this was the reason why Huyghens failed to obtain the patent he solicited for the construction of spring watches. A great number of curious contrivances at this early period in the history of Horology, may be seen in Schott's Magia Naturæ, published at Nuremberg in 1664.

Galileo was early convinced of the importance of his pendulum to the accuracy of astronomical observations; but the progress of invention is such that the steps which on looking back seem the easiest to make, are often those which are the longest delayed. Galileo recognized the principle of the isochronism of the pendulum, and recommended it as a measurer of time in 1583; yet fifty years later, although constantly using it, he had not devised a more convenient method of doing so, than is contained in the following description taken from his " Astronomical Operations."

* Utriusque Cosmi Historia. Oppenhemii, 1617.
† Huygenii Opera. Lugduni, 1724.
‡ Mémoires de l'Academie, 1717.

"A very exact time-measurer for minute intervals of time, is a heavy pendulum of any size hanged by a fine thread, which, if removed from the perpendicular and allowed to swing freely, always completes its vibrations, be they great or small, in exactly the same time."*

The mode of finding exactly by means of this the quantity of any time reduced to hours, minutes, seconds, &c., which are the divisions commonly used among astronomers, is this:—" Fit up a pendulum of any length, as for instance about a foot long, and count patiently (only for once) the number of vibrations during a natural day. Our object will be attained if we know the exact revolution of the natural day. The observer must then fix a telescope in the direction of any star, and continue to watch it till it disappears from the field of view. At that instant he must begin to count the vibrations of the pendulum, continuing all night and the following day till the return of the same star within the field of view of the telescope, and its second disappearance, as on the first night. Bearing in recollection the total number of vibrations thus made in twenty-four hours, the time corresponding to any other number of vibrations will be immediately given by the Golden Rule."

- A second extract out of Galileo's Dutch correspondence, in 1637, will show the extent of his improvements at that time:—" I come now to the second contrivance for increasing immensely the exactness of astronomical observations. I allude to my time-measurer, the precision of which is so great, and such, that it will give the exact quantity of hours, minutes, seconds, and even thirds, if their recurrence could be counted; and its constancy is such that two, four, or six such instruments will go on together so equably that one will not differ from another so much as the beat of a pulse, not only in an hour, but even in a day or a month." —
"I do not make use of a weight hanging by a thread, but a heavy and solid pendulum, made for instance of brass or copper, in the shape of a circular sector of twelve or fifteen degrees, the radius of which may be two or three palms, and the greater it is the less trouble will there be in attending it. This sector, such as I have described, I make thickest in the middle radius, tapering gradually towards the edges, where I terminate it in a tolerably sharp line, to obviate as much as possible the resistance of the air, which is the sole cause of its retardation."
—[These last words deserve notice, because, in a previous discussion, Galileo had observed that the parts of the pendulum nearest the point of suspension have a tendency to vibrate quicker than those at the other end, and seems to have thought erroneously that the stoppage of the pendulum is partly to be attributed to this cause.]
—"This is pierced in the centre, through which is passed an iron bar shaped like those on which steelyards hang, terminated below in an angle, and placed on two bronze supports, that they may wear away less during a long motion of the sector. It the sector (when accurately balanced) be removed several degrees from its perpendicular position, it will continue a reciprocal motion through a very great number of vibrations before it will stop; and in order that it may continue its motion as long as is wanted, the attendant must occasionally give it a smart push, to carry it back to large vibrations." Galileo then describes as before the method of counting the vibrations in the course of a day, and gives the rule that the lengths of two similar pendulums will have the same proportion as the squares of their times of vibration. He then continues: "Now to save the fatigue of the assistant in continually counting the vibrations, this is a convenient contrivance: A very small and delicate needle extends out from the middle of the circumference of the sector, which in passing strikes a rod fixed at one end; this rod rests upon the teeth of a wheel as light as paper, placed in a horizontal plane near the pendulum, having round it teeth cut like those of a saw, that is to say, with one side of each tooth perpendicular to the rim of the wheel and the other inclined obliquely. The rod striking against the perpendicular side of the tooth moves it, but as the same rod returns against the oblique side, it does not move it the contrary way, but slips over it and falls at the foot of the following tooth, so that the motion of the wheel will be always in the same direction. And by counting the teeth you may see at will the number of teeth passed, and consequently the number of vibrations and of particles of time elapsed. You may also fit to the axis

---

* See page 84.

H

of this first wheel a second, with a small number of teeth, touching another greater toothed wheel, &c. But it is superfluous to point out this to you, who have by you men very ingenious and well skilled in making clocks and other admirable machines ; and on this new principle, that the pendulum makes its great and small vibrations in the same time exactly, they will invent contrivances more subtle than any I can suggest; and as the error of clocks consists principally in the disability of workmen hitherto to adjust what we call the balance of the clock, so that it may vibrate regularly, my very simple pendulum, which is not liable to any alteration, affords a mean of maintaining the measures of time always equal." The contrivance thus described would be somewhat similar to the annexed representation, but it is almost certain that no such instrument was actually constructed.

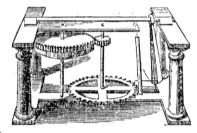

It must be owned that Galileo greatly overrated the accuracy of his timekeeper; and in asserting so positively that which he had certainly not experienced, he seems to depart from his own principles of philosophizing. It will be remarked that in this passage he still is of the erroneous opinion, that all the vibrations great or small of the same pendulum take exactly the same time ; and we have not been able to find any trace of his having ever held a different opinion, unless perhaps in the Dialogues, where he says, "If the vibrations are not exactly equal, they are at least insensibly different." This is very much at variance with the statement in the Memoirs of the Academia del Cimento, edited by their secretary Magalotti, on the credit of which Galileo's claim to the pendulum-clock chiefly rests. It is there said that experience shows that the smallest vibrations are rather the quickest, "as Galileo announced after the observation, which in 1583 he was the first to make of their approximate equality." It is not possible immediately in connexion with so glaring a misstatement, to give implicit credence to the assertion in the next sentence, that "*to obviate this inconvenience*" Galileo was the first to contrive a clock, constructed in 1649, by his son Vincenzo, in which, by the action of a weight or spring, the pendulum was constrained to move always from the same height. Indeed it appears as if Magalotti did not always tell this story in the same manner, for he is referred to as the author of the account given by Becher, " that Galileo himself made a pendulum-clock one of which was sent to Holland," plainly insinuating that Huyghens was a mere copyist.* These two accounts therefore serve to invalidate each other's credibility. Tiraboschi† asserts that, at the time he wrote, the mathematical professor at Pisa was in possession of the identical clock constructed by Treffler under Vincenzo's directions; and quotes a letter from Campani, to whom it was shown by Ferdinand, " old, rusty, and unfinished as Galileo's son made it before 1649." Viviani on the other hand says that Treffler constructed this same clock some time after Vincenzo's death (which happened in 1649), on a different principle from Vincenzo's ideas, although he says distinctly that he heard Galileo describe an application of the pendulum to a clock similar to Huyghens' contrivance. Campani did not actually see this clock till 1659, which was three years after Huyghens' invention, so that perhaps Huyghens was too easily satisfied when, on occasion of the answer which Ferdinand sent to his complaints of the Memorie del Cimento he wrote to Bouillaud, " I must however believe, since such a prince assures me, that Galileo had this idea before me."

There is another circumstance almost amounting to a proof that it was an afterthought to attribute the merit of constructing the pendulum-clock to Galileo, for on the reverse of a medal struck by Viviani, and inscribed " to the memory of his excellent instructor,"‡ is a rude exhibition of the principal objects to which Galileo's attention was directed. The pendulum is represented simply by a weight attached to a string hanging on the face of a rock. It is probable that,

---

* De nova Temporis dimetiendi ratione. Londini, 1680.
† Storia della Lett. Ital.
‡ Museum Mazuchellianum, vol. ii. Tab. cvii. p. 29.

in a design expressly intended to commemorate Galileo's inventions, Viviani would have introduced the timekeeper in the most perfect form to which it had been brought by him. Riccioli,* whose industry was unwearied in collecting every fact and argument which related in any way to the astronomical and mechanical knowledge and opinions of his time, expressly recommends swinging a pendulum, or perpendicular as it was often called (only a few years before Huyghens' publication), as much more accurate *than any clock.*† Join to all these arguments Huyghens' positive assertion, that if Galileo had conceived any such idea, he at least was entirely ignorant of it,‡ and no doubt can remain that the merit of the original invention (such as it was) rests entirely with Huyghens. The step indeed seems simple enough for a less genius than his: for the property of the pendulum was known, and the conversion of a rotatory into a reciprocating motion was known; but the connexion of the one with the other having been so long delayed, we must suppose that difficulties existed where we are not now able to perceive them, for Huyghens' improvement was received with universal admiration.

There may be many who will consider the pendulum as undeserving so long a discussion; who do not know or remember that the telescope itself has hardly done more for the precision of astronomical observations than this simple instrument, not to mention the invaluable convenience of an uniform and accurate timekeeper in the daily intercourse of life. The patience and industry of modern observers are often the theme of well-merited praise, but we must look with a still higher degree of wonder on such men as Tycho-Brahe and his contemporaries, who were driven by the want of any timekeeper on which they could depend to the most laborious expedients, and who nevertheless persevered to the best of their ability, undisgusted either by the tedium of such processes, or by the discouraging consciousness of the necessary imperfection of their most approved methods and instruments.

The invariable regularity of the pendulum's motion was soon made subservient to ulterior purposes beyond that of merely registering time. We have seen the important assistance it afforded in establishing the laws of motion; and when the theory founded on those laws was extended and improved, the pendulum was again instrumental, by a species of approximate reasoning familiar to all who are acquainted with physical inquiries, in pointing out by its minute irregularities in different parts of the earth, a corresponding change in the weight of all bodies in those different situations, supposed to be the consequence of a greater distance from the axis of the earth's rotation; since that would occasion the force of attraction to be counterbalanced by an increased centrifugal force. The theory which kept pace with the constantly increasing accuracy of such observations, proving consistent in all trials of it, has left little room for future doubts; and in this manner the pendulum in intelligent hands became the simplest instrument for ascertaining the form of the globe which we inhabit. An English astronomer, who corresponded with Kepler under the signature of Brutius (whose real name perhaps might be Bruce), had already declared his belief in 1603, that "the earth on which we tread is neither round nor globular, but more nearly of an oval figure."* There is nothing to guide us to the grounds on which he formed this opinion, which was perhaps only a lucky guess. Kepler's note upon it is: "This is not altogether to be contemned."

A farther use of the pendulum is in furnishing a general and unperishing standard of measure. This application is suggested in the third volume of the 'Reflections' of Mersenne, published in 1647, where he observes that it may be best for the future not to divide time into hours, minutes, and seconds, but to express its parts by the number of vibrations of a pendulum of given length, swinging through a given arc. It was soon seen that it would be more convenient to invert this process, and to choose as an unit of length the pendulum which should make a certain number of vibrations in the unit of time, naturally determined by the revolution of the earth on its axis. Our Royal Society took an active part in these experiments, which seem, notwithstanding their utility, to have met from the first with much of the same ridicule which was lavished

---

* Almagestum Novum, vol. i.
† Quovis horologio accuratius.
‡ Clarorum Belgarum ad Ant. Magliabech. Epistolæ. Florence, 1745, tom. i. p. 235.

* Kepleri Epistolæ.

upon them by the ignorant, when recently repeated for the same purpose. "I contend," says Graunt* in a dedication to the Royal Society, dated 1662, "against the envious schismatics of your society (who think you do nothing unless you presently transmute metals, make butter and cheese without milk, and, as their own ballad hath it, make leather without hides), by asserting the usefulness of even all your preparatory and luciferous experiments, being not the ceremonies, but the substance and principles of useful arts. For I find in trade the want of an universal measure, and have heard musicians wrangle about the just and uniform keeping of time in their consorts, and therefore cannot with patience hear that your labours about vibrations, eminently conducing to both, should be slighted, nor your pendula called swing-swangs with scorn."†

Chapter XIX.
Character of Galileo—Miscellaneous details—his Death—Conclusion.

The remaining years of Galileo's life were spent at Arcetri, where indeed, even if the Inquisition had granted his liberty, his increasing age and infirmities would probably have detained him. The rigid caution with which he had been watched in Florence was in great measure relaxed, and he was permitted to see the friends who crowded round him to express their respect and sympathy. The Grand Duke visited him frequently, and many distinguished strangers, such as Gassendi and Deodati, came into Italy solely for the purpose of testifying their admiration of his character. Among other visitors the name of Milton will be read with interest: we may probably refer to the effects of this interview the allusions to Galileo's discoveries, so frequently introduced into his poem. Milton mentions in his 'Areopagitica,' that he saw Galileo whilst in Italy, but enters into no details of his visit.

---

* Natural and Political Observations. London, 1665.
† See also Hudibras, Part II. Cant. III.
    They're guilty by their own confessions
    Of felony, and at the Sessions
    Upon the bench I will so handle 'em,
¶ ¶ ' That the vibration of this pendulum
    Shall make all taylors' yards of one
    Unanimous opinion;
    A thing he long has vaunted of,
    But now shall make it out of proof.
Hudibras was certainly written before 1663: ten years later Huyghens speaks of the idea of so employing the pendulum as a common one.

Galileo was fond of society, and his cheerful and popular manners rendered him an universal favourite among those who were admitted to his intimacy. Among these, Viviani, who formed one of his family during the three last years of his life, deserves particular notice, on account of the strong attachment and almost filial veneration with which he ever regarded his master and benefactor. His long life, which was prolonged to the completion of his 81st year in 1703, enabled him to see the triumphant establishment of the truths on account of which Galileo had endured so many insults; and even in his old age, when in his turn he had acquired a claim to the reverence of a younger generation, our Royal Society, who invited him among them in 1696, felt that the complimentary language in which they addressed him as the first mathematician of the age would have been incomplete and unsatisfactory without an allusion to the friendship that gained him the cherished title of "The last pupil of Galileo."*

Torricelli, another of Galileo's most celebrated followers, became a member of his family in October, 1641: he first learned mathematics from Castelli, and occasionally lectured for him at Rome, in which manner he was employed when Galileo, who had seen his book 'On Motion,' and augured the greatest success from such a beginning, invited him to his house—an offer which Torricelli eagerly embraced, although he enjoyed the advantages of it but for a short time. He afterwards succeeded Galileo in his situation at the court of Florence,† but survived him only a few years.

It is from the accounts of Viviani and Gherardini that we principally draw the following particulars of Galileo's person and character:—Signor Galileo was of a cheerful and pleasant countenance, especially in his old age, square built, and well proportioned in stature, and rather above the middle size. His complexion was fair and sanguine, his eyes brilliant, and his hair of a reddish cast. His constitution was naturally

---

* The words of his diploma are:—Galilæi in mathematicis disciplinis discipulus, in ærumnis socius, Italicum ingenium ita perpolivit optimis artibus ut inter mathematicos sæculi nostri facile princeps per orbem litterarium numeretur.—Tiraboschi.
† On this occasion the taste of the time showed itself in the following anagram:—
    " Evangelista Torricellieus,
    En virescit Galilæus alter."

strong, but worn out by fatigue of mind and body, so as frequently to be reduced to a state of the utmost weakness. He was subject to attacks of hypochondria, and often molested by severe and dangerous illnesses, occasioned in great measure by his sleepless nights, the whole of which he frequently spent in astronomical observations. During upwards of forty-eight years of his life, he was tormented with acute rheumatic pains, suffering particularly on any change of weather. He found himself most free from these pains whilst residing in the country, of which consequently he became very fond: besides, he used to say that in the country he had greater freedom to read the book of Nature, which lay there open before him. His library was very small, but well chosen, and open to the use of the friends whom he loved to see assembled round him, and whom he was accustomed to receive in the most hospitable manner. He ate sparingly himself; but was particularly choice in the selection of his wines, which in the latter part of his life were regularly supplied out of the Grand Duke's cellars. This taste gave an additional stimulus to his agricultural pursuits, and many of his leisure hours were spent in the cultivation and superintendence of his vineyards. It should seem that he was considered a good judge of wine; for Viviani has preserved one of his receipts in a collection of miscellaneous experiments. In it he strongly recommends that for wine of the first quality, that juice only should be employed, winch is pressed out by the mere weight of the heaped grapes, which would probably be that of the ripest fruit. The following letter, written in his 74th year, is dated, " From my prison at Arcetri.—I am forced to avail myself of your assistance and favour, agreeably to your obliging offers, in consequence of the excessive chill of the weather, and of old age, and from having drained out my grand stock of a hundred bottles, which I laid in two years ago; not to mention some minor particulars during the last two months, which I received from my Serene Master, the Most Eminent Lord Cardinal, their Highnesses the Princes, and the Most Excellent Duke of Guise, besides cleaning out two barrels of the wine of this country. Now, I beg that with all due diligence and industry, and with consideration, and taking counsel with the most refined palates, you will provide me with two cases, that is to say, with forty flasks of different wines, the most exquisite that you can find: take no thought of the expense, because I stint myself so much in all other pleasures that I can afford to lay out something at the request of Bacchus, without giving offence to his two companions Ceres and Venus. You must be careful to leave out neither Scillo nor Carino (I believe they meant to call them Scylla and Charybdis), nor the country of my master, Archimedes of Syracuse, nor Greek wines, nor clarets, &c. &c. The expense I shall easily be able to satisfy, but not the infinite obligation."

In his expenditure Galileo observed a just mean between avarice and profusion: he spared no cost necessary for the success of his many and various experiments, and spent large sums in charity and hospitality, and in assisting those in whom he discovered excellence in any art or profession, many of whom he maintained in his own house. His temper was easily ruffled, but still more easily pacified. He seldom conversed on mathematical or philosophical topics except among his intimate friends; and when such subjects were abruptly brought before him, as was often the case by the numberless visitors he was in the habit of receiving, he showed great readiness in turning the conversation into more popular channels, in such manner however that he often contrived to introduce something to satisfy the curiosity of the inquirers. His memory was uncommonly tenacious, and stored with a vast variety of old songs and stories, which he was in the constant habit of quoting and alluding to. His favourite Italian authors were Ariosto, Petrarca, and Berni, great part of whose poems he was able to repeat. His excessive admiration of Ariosto determined the side which he took against Tasso in the virulent and unnecessary controversy which has divided Italy so long on the respective merits of these two great poets; and he was accustomed to say that reading Tasso after Ariosto was like tasting cucumbers after melons. When quite a youth, he wrote a great number of critical remarks on Tasso's Gerusalemme Liberata, which one of his friends borrowed, and forgot to return. For a long time it was thought that the manuscript had perished, till the Abbé Serassi discovered it, whilst collecting materials for his Life of Tasso, published at Rome in 1785. Serassi being a violent partizan of Tasso, but also unwilling to lose the credit of the discovery, copied the manuscript, but without any intention of publishing it, "till he could find leisure for replying. properly to the sophistical and unfounded attacks of a critic so celebrated on other accounts." He announced his discovery as having been made "in one of the famous libraries at Rome," which vague indication he with some reason considered insufficient to lead to a second discovery. On Serassi's death his copy was found, containing a reference to the situation of the original; the criticisms were published, and form the greatest part of the last volume of the Milan edition of Galileo's works. The manuscript was imperfect at the time of this second discovery, several leaves having been torn out, it is not known by whom.

The opinion of the most judicious Italian critics appears to be, that it would have been more for Galileo's credit if these remarks had never been made public: they are written in a spirit of flippant violence, such as might not be extraordinary in a common juvenile critic, but which it is painful to notice from the pen of Galileo. Two or three sonnets are extant written by Galileo himself, and in two instances he has not scrupled to appropriate the conceits of the poet he affected to undervalue.* It should be mentioned that Galileo's matured taste rather receded from the violence of his early p, for at a later period of his life rejudiced to shun comparing the two; and when forced to give an opinion he said, "that Tasso's appeared the finer poem, but that Ariosto gave him the greater pleasure." Besides these sonnets, there is extant a short burlesque poem written by him, "In abuse of Gowns," when, on his first becoming Professor at Pisa, he found himself obliged by custom to wear his professional habit in every company. It is written not without humour, but does not bear comparison with Berni, whom he imitated.

There are several detached subjects treated of by Galileo, which may be noticed in this place. A letter by him containing the solution of a problem in Chances is probably the earliest notice extant of the application of mathematics to that interesting subject: the correspondence between Pascal and Fermat, with which its history is generally made to begin, not having taken place till at least twelve years later. There can be little doubt after the clear account of Carlo Dati, that Galileo was the first to examine the curve called the Cycloid, described by a point in the rim of a wheel rolling on a straight line, which he recommended as a graceful form for the arch of a bridge at Pisa. He even divined that the area contained between it and its base is exactly three times that of the generating circle. He seems to have been unable to verify this guess by strict geometrical reasoning, for Viviani tells an odd story, that in order to satisfy his doubts he cut out several large cycloids of pasteboard, but finding the weight in every trial to be rather less than three times that of the circle, he suspected the proportion to be irrational, and that there was some error in his estimation; the inquiry he abandoned was afterwards resumed with success by his pupil Torricelli.*

The account which Lagalla gives of an experiment shown in his presence by Galileo, carries the observation of the phosphorescence of the Bologna stone at least as far back as 1612.† Other writers mention the name of an alchymist, who according to them discovered it accidentally in 1603. Cesi, Lagalla, and one or two others, had passed the night at Galileo's house, with the intention of observing Venus and Saturn; but, the night being cloudy, the conversation turned on other matters, and especially on the nature of light, " on which Galileo took a small wooden box at daybreak before sunrise, and showed us some small stones in it, desiring us to observe that they were not in the least degree luminous. Having then exposed them for some time to the twilight, he shut the window again; and in the midst of the dark room showed us the stones, shining and glistening with a faint light, which we saw presently decay and become extinguished." In 1640, Liceti attempted to refer the effect of the earthshine upon the moon to a similar phosphorescent quality of that luminary, to which Galileo, then aged 76, replied by a long and able letter, enforcing the true explanation he had formerly given.

---

* Compare Son. ii. v. 8 & 9; and Son. iii. v. 2 & 3, with Ger. Lib. c. iv. st. 76, and c. vii. st. 19.—The author gladly owns his obligation for these remarks to the kindness of Sig. Panizzi, Professor of Italian in the University of London.

* Lettera di Timauro Antiate. Firenze, 1663.
† De phænomenis in orbe Lunæ. Venetiis, 1612.

Although quite blind, and nearly deaf, the intellectual powers of Galileo remained to the end of his life; but he occasionally felt that he was overworking himself, and used to complain to his friend Micanzio that he found his head too busy for his body. "I cannot keep my restless brain from grinding on, although with great loss of time; for whatever idea comes into my head with respect to any novelty, drives out of it whatever I had been thinking of just before." He was busily engaged in considering the nature of the force of percussion, and Torricelli was employed in arranging his investigations for a continuation of the 'Dialogues on Motion,' when he was seized with an attack of fever and palpitation of the heart, which, after an illness of two months, put an end to his long, laborious, and useful life, on the 8th of January, 1642, just one year before his great successor Newton was born.

The malice of his enemies was scarcely allayed by his death. His right of making a will was disputed, as having died a prisoner to the Inquisition, as well as his right to burial in consecrated ground. These were at last conceded, but Urban anxiously interfered to prevent the design of erecting a monument to him in the church of Santa Croce, in Florence, for which a large sum had been subscribed. His body was accordingly buried in an obscure corner of the church, which for upwards of thirty years after his death was unmarked even by an inscription to his memory. It was not till a century later that the splendid monument was erected which now covers his and Viviani's remains. When their bodies were disinterred in 1737 for the purpose of being removed to their new resting-place, Capponi, the president of the Florentine Academy, in a spirit of spurious admiration, mutilated Galileo's body, by removing the thumb and fore-finger of the right-hand, and one of the vertebræ of the back, which are still preserved in some of the Italian museums. The monument was put up at the expense of his biographer, Nelli, to whom Viviani's property descended, charged with the condition of erecting it. Nor was this the only public testimony which Viviani gave of his attachment. The medal which he struck in honour of Galileo has already been mentioned; he also, as soon as it was safe to do so, covered every side of the house in which he lived with laudatory inscriptions to the same effect. A bust of Galileo was placed over the door, and two bas-reliefs on each side representing some of his principal discoveries. Not less than five other medals were struck in honour of him during his residence at Padua and Florence, which are all engraved in Venturi's Memoirs.

There are several good portraits of Galileo extant, two of which, by Titi and Subtermanns, are engraved in Nelli's Life of Galileo. Another by Subtermanns is in the Florentine Gallery, and an engraving from a copy of this is given by Venturi. There is also a very fine engraving from the original picture. An engraving from another original picture is in the frontispiece of the Padua edition of his works. Salusbury seems in the following passage to describe a portrait of Galileo painted by himself: "He did not contemn the other inferior arts, for he had a good hand in sculpture and carving; but his particular care was to paint well. By the pencil he described what his telescope discovered; in one he exceeded art, in the other, nature. Osorius, the eloquent bishop of Sylva, esteems one piece of Mendoza the wise Spanish minister's felicity, to have been this, that he was contemporary to Titian, and that by his hand he was drawn in a fair tablet. And Galilæus, lest he should want the same good fortune, made so great a progress in this curious art, that he became his own *Buonarota;* and because there was no other copy worthy of his pencil, drew himself." No other author makes the slightest allusion to such a painting; and it appears more likely that Salusbury should be mistaken than that so interesting a portrait should have been entirely lost sight of.

Galileo's house at Arcetri was standing in 1821, when Venturi visited it, and found it in the same state in which Galileo might be supposed to have left it. It is situated nearly a mile from Florence, on the south-eastern side, and about a gun-shot to the north-west of the convent of St. Matthew. Nelli placed a suitable inscription over the door of the house, which belonged in 1821 to a Signor Alimari.*

Although Nelli's Life of Galileo disappointed the expectations that had been formed of it, it is impossible for any admirer of Galileo not to feel the greatest degree of gratitude towards

---

* Venturi.

him, for the successful activity with which he rescued so many records of the illustrious philosopher from destruction. After Galileo's death, the principal part of his books, manuscripts, and instruments, were put into the charge of Viviani, who was himself at that time an object of great suspicion; most of them he thought it prudent to conceal, till the superstitious outcries against Galileo should be silenced. At Viviani's death, he left his library, containing a very complete collection of the works of all the mathematicians who had preceded him (and amongst them those of Galileo, Torricelli, and Castelli, all which were enriched with notes and additions by himself), to the hospital of St. Mary at Florence, where an extensive library already existed. The directors of the hospital sold this unique collection in 1781, when it became entirely dispersed. The manuscripts in Viviani's possession passed to his nephew, the Abbé Panzanini, together with the portraits of the chief personages of the Galilean school, Galileo's instruments, and, among other curiosities, the emerald ring which he wore as a member of the Lyncean Academy. A great number of these books and manuscripts were purchased at different times by Nelli, after the death of Panzanini, from his relations, who were ignorant or regardless of their value. One of his chief acquisitions was made by an extraordinary accident, related by Tozzetti with the following details, which we repeat, as they seem to authenticate the story:—"In the spring of 1739, the famous Doctor Lami went out according to his custom to breakfast with some of his friends at the inn of the Bridge, by the starting-place; and as he and Sig. Nelli were passing through the market, it occurred to them to buy some Bologna sausages from the pork-butcher, Cioci, who was supposed to excel in making them. They went into the shop, had their sausages cut off and rolled in paper, which Nelli put into his hat. On reaching the inn, and calling for a plate to put them in, Nelli observed that the paper in which they had been rolled was one of Galileo's letters. He cleaned it as well as he could with his napkin, and put it into his pocket without saying a word to Lami; and as soon as he returned into the city, and could get clear of him, he flew to the shop of Cioci, who told him that a servant whom he did not know brought him from time to time. similar letters, which he bought by weight as waste paper. Nelli bought all that remained, and on the servant's next reappearance in a few days, he learned the quarter whence they came, and after some time succeeded at a small expense in getting into his own possession an old corn-chest, containing all that still remained of the precious treasures which Viviani had concealed in it ninety years before."*

The earliest biographical notice of Galileo is that in the Obituary of the Mercurio Italico, published at Venice in 1647, by Vittorio Siri. It is very short, but contains an exact enumeration of his principal works and discoveries. Rossi, who wrote under the name of Janus Nicius Erythræus, introduced an account of Galileo in his Pinacotheca Imaginum Illustrium, in which the story of his illegitimacy first made its appearance. In 1664, Salusbury published a life of Galileo in the second volume of his Mathematical Collections, the greater part of which is a translation of Galileo's principal works. Almost the whole edition of the second volume of Salusbury's book was burnt in the great fire of London. Chauffepié says that only one copy is known to be extant in England: this is now in the well-known library of the Earl of Macclesfield, to whose kindness the author is much indebted for the use he has been allowed to make of this unique volume. A fragment of this second volume is in the Bodleian Library at Oxford. The translations in the preceding pages are mostly founded upon Salusbury's version. Salusbury's account, although that of an enthusiastic admirer of Galileo, is too prolix to be interesting: the general style of the performance may be guessed from the title of the first chapter—' Of Man, in general, and how he excelleth all the other Animals.' After informing his readers that Galileo was born at Pisa, he proceeds:—" Italy is affirmed to have been the first that peopled the world after the universal deluge, being governed by Janus, Cameses, and Saturn, &c." His description of Galileo's childhood is somewhat quaint. " Before others had left making of dirt pyes, he was framing of diagrams; and whilst others were whipping of toppes, he was considering the cause of their motion." It is on the

* Notizie sul Ingrandimento delle Scienze Fisiche. Firenze, 1780.

whole tolerably correct, especially if we take into account that Salusbury had not yet seen Viviani's Life, though composed some years earlier.

The Life of Galileo by Viviani was first written as an outline of an intended larger work, but this latter was never completed. This sketch was published in the Memoirs of the Florentine Academy, of which Galileo had been one of the annual presidents, and afterwards prefixed to the complete editions of Galileo's works; it is written in a very agreeable and flowing style, and has been the groundwork of most subsequent accounts. Another original memoir by Niccolò Gherardini, was published by Tozzetti. A great number of references to authors who have treated of Galileo is given by Sach in his Onomasticon. An approved Latin memoir by Brenna is in the first volume of Fabroni's Vitæ. Italorum Illustrium; he has however fallen into several errors: this same work contains the lives of several of his principal followers.

The article in Chauffepié's Continuation of Bayle's Dictionary does not contain anything which is not in the earlier accounts.

Andrès wrote an essay entitled 'Saggio sulla Filosofia del Galileo,' published at Mantua 1776; and Jagemann published his 'Geschichte des Leben des Galileo' at Leipzig, in 1787;* neither of these the author has been able to meet with. An analysis of the latter may be seen in Kästner's ' Geschichte der Mathematik, Göttingen, 1800,' from which it does not appear to contain any additional details. The ' Elogio del Galileo' by Paolo Frisi, first published at Leghorn in 1775, is, as its title expresses, rather in the nature of a panegyric than of a continuous biographical account. It is written with very great elegance and intimate knowledge of the subjects of which it treats. Nelli gave several curious particulars with respect to Galileo in his ' Saggio di Storia Letteraria Fiorentina, Lucca, 1759;' and in 1793 published his large work entitled 'Vita e Commercio Letterario di Galileo Galilei.' So uninteresting a book was probably never written from such excellent materials. Two thick quarto volumes are filled with repetitions of the accounts that were already in print, the bulky preparation of which compelled the author to forego the publication of the vast collection of original documents which his unwearied zeal and industry had collected. This defect has been in great measure supplied by Venturi in 1818 and 1821, who has not only incorporated in his work many of Nelli's manuscripts, but has brought together a number of scattered notices of Galileo and his writings from a variety of outlying sources—a service which the writer is able to appreciate from having gone through the greatest part of the same labour before he was fortunate enough to meet with Venturi's book. Still there are many letters cited by Nelli, which do not appear either in his book or Venturi's. Carlo Dati, in 1663, quotes " the registers of Galileo's correspondence arranged in alphabetical order, in ten large volumes."* The writer has no means of ascertaining what collection this may have been; it is difficult to suppose that one so arranged should have been lost sight of. It is understood that a life of Galileo is preparing at this moment in Florence, by desire of the present Grand Duke, which will probably throw much additional light on the character and merits of this great and useful philosopher.

The first editions of his various treatises, as mentioned by Nelli, are given below. Clement, in his ' Bibliothèque Curieuse,' has pointed out such among them, and the many others which have been printed, as have become rare.

The Florentine edition is the one used by the Academia della Crusca for their references; for which reason its paging is marked in the margin of the edition of Padua, which is much more complete, and is the one which has been on the present occasion principally consulted.

The latter contains the Dialogue on the System, which was not suffered to be printed in the former editions. The twelve first volumes of the last edition of Milan are a mere transcript of that of Padua: the thirteenth contains in addition the Letter to the Grand Duchess, the Commentary on Tasso, with some minor pieces. A complete edition is still wanted, embodying all the recently discovered documents, and omitting the verbose commentaries, which, however useful when they were written, now convey little information that cannot be more agreeably and more profitably learned in treatises of a later date.

---

* Venturi.

* Lettera di Timauro Antiate.

Such was the life, and such were the pursuits, of this extraordinary man. The numberless inventions of his acute industry; the use of the telescope, and the brilliant discoveries to which it led; the patient investigation of the laws of weight and motion; must all be looked upon as forming but a part of his real merits, as merely particular demonstrations of the spirit in which he everywhere withstood the despotism of ignorance, and appealed boldly from traditional opinions to the judgments of reason and common sense. He claimed and bequeathed to us the right of exercising our faculties in examining the beautiful creation which surrounds us. Idolized by his friends, he deserved their affection by numberless acts of kindness; by his good humour, his affability, and by the benevolent generosity with which he devoted himself and a great part of his limited income to advance their talents and fortunes. If an intense desire of being useful is everywhere worthy of honour; if its value is immeasurably increased, when united to genius of the highest order; if we feel for one who, notwithstanding such titles to regard, is harassed by cruel persecution,—then none deserve our sympathy, our admiration, and our gratitude, more than Galileo.

List of Galileo's Works.

Le Operazioni del Compasso Geom. e Milit.	Padova, 1606.	Fol.
Difesa di Gal. Galilei contr. all. cal. et impost. di Bald. Capra	Venezza, 1607.	4to.
Sydereus Nuncius	Venetiis, 1610.	4to.
Discorso int. alle cose che stanno in su l'Acqua	Firenze, 1612.	4to.
Novantiqua SS. PP. Doctrina de S. Scripturæ Testimoniis	Argent, 1612.	4to.
Istoria e Demostr. int. alle Macchie Solari	Roma, 1613.	4to.
Risp. alle oppos. del S. Lod. delle Colombe e del S. Vinc. di Grazia	Firenze, 1615.	4to.
Discorso delle Comete di Mario Guiducci	Firenze, 1619.	4to.
Dialogo sopra i due Massimi Sistemi del Mondo	Firenze, 1632.	4to,
Discorso e Demostr. intorno alle due nuove Scienze	Leida, 1638.	4to.
Della Scienza Meccanica	Ravenna, 1649.	4to.
Trattato della Sfera	Roma, 1655.	4to.
Discorso sopra il Flusso e Reflusso. (Scienze Fisiche di Tozzetti.)	Firenze, 1780.	4to.
Considerazioni sul Tasso	Roma, 1793.	
Trattato della Fortificazione. (Memorie di Venturi.)	Modena, 1818.	4to.

The editions of his collected works (in which is contained much that was never published separately) are—

Opere di Gal. Galilei, Linc. Nob. Fior. &c.	Bologna, 1656.	2 vols. 4to.
Opere di Gal. Galilei, Nob. Fior. Accad. Linc. &c.	Firenze, 1718.	3 vols. 4to.
Opere di Gal. Galilei	Padova, 1744.	4 vols. 4to.
Opere di Gal. Galilei	Milano, 1811.	13 vols. 8vo.

CORRECTIONS.

Page Co . Line.
5   1   2, *Add*: His instructor was the celebrated botanist, Andreas Cæsalpinus, who was professor of medicine at Pisa from 1567 to 1592. Hist. Acad. Pisan.; Pisis, 1791.
8   2   18, *Add*: According to Kästner, his German name was Wursteisen.
8   2   21, *for* 1588 *read* 1586.
15 1   57, *for* 1632 read 1630.
17 1   29. Salusbury alludes to the instrument described and figured in "The Use of the Sector, Crosse Staffe, and other Instruments. London, 1624." It is exactly Galileo's Compass.
17 1   52, *for* Burg, a German, *read* Burgi, a Swiss.
27 2   17. The author here called Brutti was an Englishman: his real name, perhaps, was Bruce. See p. 99.
50 1   14. Kepler's Epitome was not published till 1619: it was then inserted in the Index.
73 1   60, *for* under *read* turned from.
80 1   50, *for* any *read* an indefinitely small.

LIFE OF KEPLER.

CHAPTER I.

Introduction—Birth and Education of Kepler—He is appointed Astronomical Professor at Gratz—Publishes the 'Mysterium Cosmographicum.'

IN the account of the life and discoveries of Galileo, we have endeavoured to inculcate the safety and fruitfulness of the method followed by that great reformer in his search after physical truth. As his success furnishes the best instance of the value of the inductive process, so the failures and blunders of his adversaries supply equally good examples of the dangers and the barrenness of the opposite course. The history of JOHN KEPLER might, at the first view, suggest conclusions somewhat inconsistent with this remark. Every one who is but moderately acquainted with astronomy is familiar with the discoveries which that science owes to him; the manner in which he made them is, perhaps, not so generally known. This extraordinary man pursued, almost invariably, the hypothetical method. His life was passed in speculating on the results of a few principles assumed by him, from very precarious analogies, as the causes of the phenomena actually observed in Nature. We nevertheless find that he did, in spite of this unphilosophical method, arrive at discoveries which have served as guides to some of the most valuable truths of modern science.

The difficulty will disappear if we attend more closely to the details of Kepler's investigations. We shall perceive that to an unusual degree of rashness in the formation of his systems, he added a quality very rarely possessed by philosophers of the hypothetical school. One of the greatest intellectual vices of the latter was a wilful blindness to the discrepancy of facts from their creed, a perverse and obstinate resistance to physical evidence, leading not unfrequently to an attempt at disguising the truth. From this besetting sin of the school, which from an intellectual fault often degenerated into a moral one, Kepler was absolutely free. Scheme after scheme, resting originally upon little beyond his own glowing imagination, but examined and endeared by the ceaseless labour of years, was unhesitatingly sacrificed, as soon as its insufficiency became indisputable, to make room for others as little deserving support. The history of philosophy affords no more remarkable instance of sincere uncompromising love of truth. To this virtue he owed his great discoveries: it must be attributed to his unhappy method that he made no more.

In considering this opinion upon the real nature of Kepler's title to fame, it ought not to be forgotten that he has exposed himself at a disadvantage on which certainly very few philosophers would venture. His singular candour allowed him to comment upon his own errors with the same freedom as if scrutinizing the work of a stranger; careless whether the impression on his readers were favourable or otherwise to himself, provided it was instructive. Few writers have spoken so much, and so freely of themselves, as Kepler. He records, on almost every occasion, the train of thought by which he was led to each of the discoveries that eventually repaid his perseverance; and he has thus given us a most curious and interesting view of the workings of a mind of great, though eccentric power. "In what follows," says he (when introducing a long string of suppositions, of which he had already discovered the fallacy), "let the reader pardon my credulity, whilst working out all these matters by my own ingenuity. For it is my opinion that the occasions by which men have acquired a knowledge of celestial phenomena are not less admirable than the discoveries themselves.'" Agreeing altogether with this opinion in its widest application, we have not scrupled, in the following sketch, to introduce at some length an account even of Kepler's erroneous speculations; they are in themselves very amusing, and will have the additional utility of proving the dangerous tendency of his method; they will show by how many absurd theories, and how many years of wasted labour, his real discoveries and services to science lie surrounded.

JOHN KEPLER was born (as we are assured by his earliest biographer Hantsch) in long. 29° 7′, lat. 48° 54′, on the 21st day of December, 1571. On this spot stands the imperial city of Weil, in the duchy of Wirtemberg. His parents were Henry Kepler and Catherine Guldenmann, both of noble, though decayed families. Henry Kepler, at the time of his marriage, was a petty officer in the Duke of Wirtemberg's service; and a few years after the birth of his eldest son John, he joined the army then serving in the Netherlands. His wife followed him, leaving their son, then in his fifth year, at Leonberg, under the care of his grandfather. He was a seven months child, very weak and sickly; and after recovering with difficulty from a severe attack of small-pox, he was sent to school in 1577. Henry Kepler's limited income was still farther reduced on his return into Germany, the following year, in consequence of the absconding of one of his acquaintance, for whom he had incautiously become surety. His circumstances were so much narrowed by this misfortune, that he was obliged to sell his house, and nearly all that he possessed, and for several years he supported his family by keeping a tavern at Elmendingen. This occasioned great interruption to young Kepler's education; he was taken from school, and employed in menial services till his twelfth year, when he was again placed in the school at Elmendingen. In the following year he was again seized with a violent illness, so that his life was almost despaired of. In 1586, he was admitted into the monastic school of Maulbronn, where the cost of his education was defrayed by the Duke of Wirtemberg. This school was one of those established on the suppression of the monasteries at the Reformation, and the usual course of education followed there required that the students, after remaining a year in the superior classes, should offer themselves for examination at the college of Tubingen for the degree of bachelor: they then returned to their school with the title of veterans; and after completing the studies taught there, they were admitted as resident students at Tubingen, proceeded in about a year to the degree of master, and were then allowed to commence their course of theology. The three years of Kepler's life following his admission to Maulbronn, were marked by periodical returns of several of the disorders which had well nigh proved fatal to him in his childhood. During the same time disagreements arose between his parents, in consequence of which his father quitted his home, and soon after died abroad. After his father's departure, his mother also quarrelled with her relations, having been treated, says Hantsch, "with a degree of barbarity by her husband and brother-in-law that was hardly exceeded even by her own perverseness:" one of his brothers died, and the family-affairs were in the greatest confusion. Notwithstanding these disadvantages, Kepler took his degree of master in August 1591, attaining the second place in the annual examination. The first name on the list was John Hippolytus Brentius.

Whilst he was thus engaged at Tubingen, the astronomical lectureship at Gratz, the chief town of Styria, became vacant by the death of George Stadt, and the situation was offered to Kepler. Of this first occasion of turning his thoughts towards astronomy, he has himself given the following account: "As soon as I was of an age to feel the charms of philosophy, I embraced every part of it with intense desire, but paid no especial regard to astronomy. I had indeed capacity enough for it, and learned without difficulty the geometrical and astronomical theorems occurring in the usual course of the school, being well grounded in figures, numbers, and proportions. But those were compulsory studies—there was nothing to show a particular turn for astronomy. I was educated at the expense of the Duke of Wirtemberg, and when I saw such of my companions as the duke selected to send abroad shrink in various ways from their employments, out of fondness for home, I, who was more callous, had early made up my mind to go with the utmost readiness whithersoever I might be sent. The first offering itself was an astronomical post, which I was in fact forced to accept by the authority of my tutors; not that I was alarmed, in the manner I had condemned in others, by the remoteness of the situation, but by the unexpected and contemptible nature of the office, and by the slightness of my information in this branch of philosophy. I entered on it, therefore, better furnished with talent than knowledge: with many protestations that I was not abandoning my claim to be provided for in some other more brilliant profession. What progress I made in the first two years of my studies, may be seen in my 'Mysterium Cosmographicum;' and the encouragement given me by my tutor, Mästlin, to take up the science of astronomy, may be read in the same book, and in his letter which is prefixed to the 'Narrative of Rheticus.' I looked on that discovery as of the highest importance, and still more so, because I saw how greatly it was approved by Mästlin."

The nature of the singular work to which Kepler thus refers with so much complacency, will be best shown by quoting some of the most remarkable parts of it, and especially the preface, in which he briefly details some of the theories he successively examined and rejected, before detecting (as he imagined he had here done) the true cause of the number and order of the heavenly bodies. The other branches of philosophy with which he occupied himself in his younger years, were those treated by Scaliger in his 'Exoteric Exercises,' to the study of which book Kepler attributed the formation of many of his opinions; and he tells us that he devoted much time " to the examination of the nature of heaven, of souls, of genii, of the elements, of the essence of fire, of the cause of fountains, the ebb and flow of the tide, the shape of the continents, and inland seas, and things of this sort." He also says, that by his first success with the heavens, his hopes were greatly inflamed of discovering similar analogies in the rest of the visible world, and for this reason, named his book merely a Prodromus, or Forerunner, meaning, at some future period, to subjoin the Aftercomer, or Sequel. But this intention was never fulfilled; either his imagination failed him, or, what is more likely, the laborious calculations in which his astronomical theories engaged him, left him little time for turning his attention to objects unconnected with his first pursuit.

It is seldom that we are admitted to trace the progress of thought in those who have distinguished themselves by talent and originality; and although the whole of the following speculations begin and end in error, yet they are so characteristic, and exhibit such an extraordinary picture of the extravagances into which Kepler's lively imagination was continually hurrying him, that we cannot refrain from citing nearly the whole preface. From it, better than from any enumeration of peculiarities, the reader will at once apprehend the nature of his disposition.

"When I was attending the celebrated Mästlin, six years ago, at Tubingen, I was disturbed by the manifold inconveniences of the common theory of the universe, and so delighted with Copernicus, whom Mästlin was frequently in the habit of quoting with great respect, that I not only often defended his propositions in the physical disputations of the candidates, but also wrote a correct essay on the primary motion, maintaining, that it is caused by the rotation of the earth. And I was then at that point that I attributed to the earth the motion of the sun on physical (or, if you will, on metaphysical) grounds, as Copernicus had done for mathematical reasons. And, by this practice, I came by degrees, partly from Mästlin's instructions, and partly from my own efforts, to understand the superior mathematical convenience of the system of Copernicus beyond Ptolemy's. This labour might have been spared me, by Joachim Rheticus, who has shortly and clearly explained everything in his first Narrative. While incidentally engaged in these labours, in the intermission of my theology, it happened conveniently that I succeeded George Stadt in his situation at Gratz, where the nature of my office connected me more closely with these studies. Everything I had learned from Mästlin, or had acquired of myself, was there of great service to me in explaining the first elements of astronomy. And, as in Virgil, ' Fama mobilitate viget, viresque acquirit eundo,' so it was with me, that the diligent thought on these things was the occasion of still further thinking: until, at last, in the year 1595, when I had some intermission of my lectures allowed me, I brooded with the whole energy of my mind on this subject. There were three things in particular, of which I pertinaciously sought the causes why they are not other than they are : the number, the size, and the motion of the orbits. I attempted the thing at first with numbers, and considered whether one of the orbits might be double, triple, quadruple, or any other multiple of the others, and how much, according to Copernicus, each differed from the rest. "I spent a great deal of time in that labour, as if it were mere sport, but could find no equality either in the proportions or the differences, and I gained nothing from this beyond imprinting deeply in my memory the distances as assigned by Copernicus; unless, perhaps, reader, this record of my various attempts may force your assent, backwards and forwards, as the waves of the sea; until tired at length, you will willingly repose yourself, as in a safe haven, on the reasons explained in this book. However, I was comforted in some degree, and my hopes of success were supported as well by other reasons which will follow presently, as by observing that the motions in every case seemed to be connected with the distances, and that where there was a great gap between the orbits, there was the same between the motions. And I reasoned, that if God had adapted motions to the orbits in some relation to the distances, it was probable that he had also arrayed the distances themselves in relation to something else.

"Finding no success by this method, I tried another, of singular audacity. I inserted a new planet between Mars and Jupiter, and another between Venus and Mercury, both of which I supposed invisible, perhaps on account of their smallness, and I attributed to each a certain period of revolution.* I thought that I could thus contrive some equality of proportions, increasing between every two, from the sun to the fixed stars. For instance, the Earth is nearer Venus in parts of the terrestrial orbit, than Mars is to the Earth in parts of the orbit of Mars. But not even the interposition of a new planet sufficed for the enormous gap between Mars and Jupiter; for the proportion of Jupiter to the new planet was still greater than that of Saturn to Jupiter. And although, by this supposition, I got some sort of a proportion, yet there was no reasonable conclusion, no certain determination of the number of the planets either towards the fixed stars, till we should get as far as them, nor ever towards the Sun, because the division in this proportion of the residuary space within Mercury might be continued without end. Nor could I form any conjecture, from the mobility of particular numbers, why, among an infinite number, so few should be moveable. The opinion advanced by Rheticus in his Narrative is improbable, where he reasons from the sanctity of the number six to the number of the six moveable heavens; for he who is inquiring of the frame of the world itself, must not derive reasons from these numbers, which have gained importance from things of later date.

"I sought again, in another way, whether the distance of every planet is not as the residuum of a sine; and its motion as the residuum of the sine of the complement in the same quadrant.

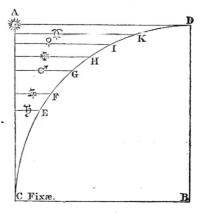

"Conceive the square A B to be constructed, whose side A C is equal to the semidiameter of the universe. From the angle B opposite to A the place of the sun, or centre of the world, describe the quadrant D C with the radius B C. Then in A C, the true radius of the world, let the sun, fixed stars, and planets be marked at their respective distances, and from these points draw lines parallel to B C, meeting the quadrant. I imagined the moving force acting on each of the planets to be in the proportion of these parallels. In the line of the sun is infinity, because A D is touched, and not cut, by the quadrant: therefore the moving force is infinite in the sun, as deriving no motion except from its own act. In Mercury the infinite line is cut off at K, and therefore at this point the motion is comparable with the others. In the fixed stars the line is altogether lost, and compressed into a mere point C; therefore at that point there is no moving force. This was the theorem, which was to be tried by cal-

---

* The following scrupulous note added by Kepler in 1621 to a subsequent edition of this work, deserves to be quoted. It shows how entirely superior he was to the paltriness of attempting to appropriate the discoveries of others, of which many of his contemporaries had exhibited instances even on slighter pretences than this passage might have afforded him. The note is as follows: "Not circulating round Jupiter like the Medicæan stars. Be not deceived. I never had them in my thoughts, but, like the other primary planets, including the sun in the centre of the system within their orbits."

culation; but if any one will reflect that two things were wanting to me, first, that I did not know the size of the *Sinus Totus*, that is, the radius of the proposed quadrant; secondly, that the energies of the motions were not thus expressed otherwise than in relatiou one to another; whoever, I say, well considers this, will doubt, not without reason, as to the progress I was likely to make in this difficult course. And yet, with unremitting labour, and an infinite reciprocation of sines and arcs, I did get so far as to be convinced that this theory could not hold.

"Almost the whole summer was lost in these annoying labours; at last, by a trifling accident, I lighted more nearly on the truth. I looked on it as an interposition of Providence, that I should obtain by chance, what I had failed to discover with my utmost exertions; and I believed this the more, because I prayed constantly that I might succeed, if Copernicus had really spoken the truth. It happened on the 9th or 19th* day of July, in the year 1595, that, having occasion to show, in my lecture-room, the passages of the great conjunctions through eight signs, and how they pass gradually from one trine aspect to another, I inscribed in a circle

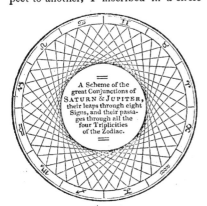

A Scheme of the great Conjunctions of SATURN & JUPITER, their leaps through eight Signs, and their passages through all the four Triplicities of the Zodiac.

a great number of triangles, or quasi-triangles, so that the end of one was made the beginning of another. In this manner a smaller circle was shadowed out by the points in which the lines crossed each other.

"The radius of a circle inscribed in a triangle is half the radius of that described about it; therefore the pro-

---

* This inconvenient mode of dating was necessary before the new or Gregorian style was universally adopted.

portion between these two circles struck the eye as almost identical with that between Saturn and Jupiter, and the triangle is the first figure, just as Saturn and Jupiter are the first planets. On the spot I tried the second distance between Jupiter and Mars with a square, the third with a pentagon, the fourth with a hexagon. And as the eye again cried out against the second distance between Jupiter and Mars, I combined the square with a triangle and a pentagon. There would be no end of mentioning every trial. The failure of this fruitless attempt was the beginning of the last fortunate one; for I reflected, that in this way I should never reach the sun, if I wished to observe the same rule throughout; nor should I have any reason why there were six, rather than twenty or a hundred moveable orbits. And yet figures pleased me, as being quantities, and as having existed before the heavens; for quantity was created with matter, and the heavens afterwards. But if (this was the current of my thoughts), in relation to the quantity and proportion of the six orbits, as Copernicus has determined them among the infinite other figures, five only could be found having peculiar properties above the rest, my business would be done. And then again it struck me, what have plane figures to do among solid orbits? Solid bodies ought rather to be introduced. This, reader, is the invention and the whole substance of this little work; for if any one, though but moderately skilled in geometry, should hear these words hinted, the five regular solids will directly occur to him with the proportions of their circumscribed and inscribed spheres: he has immediately before his eyes that scholium of Euclid to the 18th proposition of his 13th Book, in which it is proved to be impossible that there should be, or be imagined, more than five regular bodies.

"What is worthy of admiration (since I had then no proof of any prerogatives of the bodies with regard to their order) is, that employing a conjecture which was far from being subtle, derived from the distances of the planets, I should at once attain my end so happily in arranging them, that I was not able to change anything afterwards with the utmost exercise of my reasoning powers. In memory of the event, I write down here for you the sentence, just as it fell from me, and in the words in which it was that moment conceived:—The Earth is the circle, the measurer of all; round it describe a dodecahedron, the circle including this will be Mars. Round Mars describe a tetrahedron, the circle including this will be Jupiter. Describe a cube round Jupiter, the circle including this will be Saturn. Now, inscribe in the Earth an icosaedron, the circle inscribed in it will be Venus. Inscribe an octaedron in Venus, the circle inscribed in it will be Mercury. This is the reason of the number of the planets.

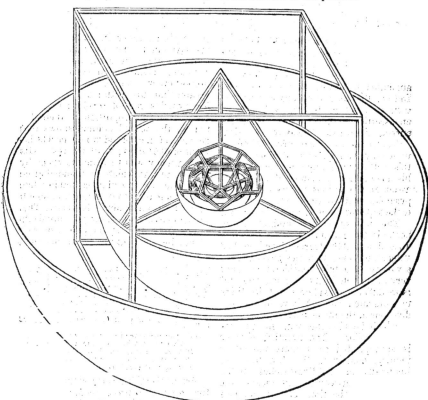

"This was the cause, and such the success, of my labour: now read my propositions in this book." The intense pleasure I have received from this discovery never can be told in words. I regretted no more the time wasted; I tired of no labour; I shunned no toil of reckoning; days and nights I spent in calculations; until I could see whether this opinion would agree with the orbits of Copernicus, or whether my joy was to vanish into air. I willingly subjoin that sentiment of Archytas, as given by Cicero:— 'If I could mount up into heaven, and thoroughly perceive the nature of the world, and beauty of the stars, that admiration would be without a charm for me; unless I had some one like you, reader, candid, attentive,' and eager for knowledge, to whom to describe it.' If you acknowledge this feeling, and are candid, you will refrain from blame, such as not without cause I anticipate; but if, leaving that to itself, you fear lest these things be not ascertained, and that I have shouted triumph before victory, at least approach these pages, and learn the matter in consideration: you will not find, as just now, new and unknown planets interposed; that boldness of mine is not approved, but those old ones very little loosened, and so furnished by the interposition (however absurd you may think it) of rectilinear figures, that in future you may give a reason to the rustics when they ask for the hooks which keep the skies from falling.— Farewell."

· In the third chapter Kepler mentions, that a thickness must be allowed to each orb sufficient to include the greatest and least distance of the planet from the sun. The form and result of his comparison with the real distances are as follows:—

If the inner surface of the orbit of		be taken at 1000, then the outer one of		According to Copernicus they are	Book V.
	Saturn		Jupiter = 577		635 Ch. 9
	Jupiter		Mars = 333		333 — 14
	Mars		Earth = 795		757 — 19
	Earth		Venus = 795		794 — 21, 22
	Venus		Mercury = 577		723 — 27

It will be observed, that Kepler's results were far from being entirely satisfactory; but he seems to have flattered himself, that the differences might be attributed to erroneous measurements. Indeed, the science of observation was then so much in its infancy, that such an assertion might be made without incurring much risk of decisive refutation.

Kepler next endeavoured to determine why the regular solids followed in this rather than any other order; and his imagination soon created a variety of essential distinctions between the cube, pyramid, and dodecahedron, belonging to the superior planets, and the other two.

The next question examined in the book, is the reason why the zodiac is divided into 360 degrees; and on this subject, he soon becomes enveloped in a variety of subtle considerations, (not very intelligible in the original, and still more difficult to explain shortly to others unacquainted with it,) in relation to the divisions of the musical scale; the origin of which he identifies with his five favourite solids. The twentieth chapter is appropriated to a more interesting inquiry, containing the first traces of his finally successful researches into the proportion between the distances of the planets, and the times of their motions round the sun. He begins with the generally admitted fact, that the more distant planets move more slowly; but in order to show that the proportion, whatever it may be, is not the simple one of the distances, he exhibits the following little Table:—

	♄ D. Scr.	♃ D. Scr.	♂ D.Scr.	⊕ D.Scr.	♀ D.Scr.	☿ D.Scr.
♄	10759.12					
♃	6159	4332.37				
♂	1785	1282	686.59			
⊕	1174	843	452	365.15		
♀	844	606	325	262.30	224.42	
☿	434	312	167	135	115	87.58

At the head of each vertical column is placed the real time (in days and sexagesimal parts) of the revolution of the planet placed above it, and underneath the days due to the other inferior planets, if they observed the proportion of distance. Hence it appears that this proportion in every case gives a time greater than the truth; as for instance, if the earth's rate of revolution were to Jupiter's in the proportion of their distances, the second column shows that the time of her period would be 843 instead of $365\frac{1}{4}$ days; so of the rest. His next attempt was to compare them by two by two, in which he found that he arrived at a proportion something like the proportion of the distances, although as yet far from obtaining it exactly. This process amounts to taking the quotients obtained by dividing the period of each planet by the period of the one next beyond.

For if each of the Periods of		be successively taken to consist of 1000 equal parts, the periods of the planet next below will contain of those parts in	
	♄ 10759.27		♃ 403
	♃ 4332.37		♂ 159
	♂ 686.59		⊕ 532
	⊕ 365.15		♀ 615
	♀ 244.42		☿ 392

But if the distance of each planet in succession be taken to consist of 1000 equal parts, the distance of the next below will contain, according to Copernicus, in		
	♃	572
	♂	290
	⊕	658
	♀	719
	☿	500

From this table he argued that to make the proportions agree, we must assume one of two things, "either that the moving intelligences of the planets are weakest in those which are farthest from the Sun, or that there is one moving intelligence in the Sun, the common centre forcing them all round, but those most violently which are nearest, and that it languishes in some sort, and grows weaker at the most distant, because of the remoteness and the attenuation of the virtue."

We stop here to insert a note added by Kepler to the later editions, and shall take advantage of the same interruption to warn the reader not to confound this notion of Kepler with the theory of a gravitating force towards the Sun, in the sense in which we now use those words. According to our theory, the effect of the presence of the Sun upon the planet is to pull it towards the centre in a straight line, and the effect of the motion thus produced combined with the motion of the planet, which if undisturbed would be in a straight line inclined to the direction of the radius, is, that it describes a curve round the Sun. Kepler considered his planets as perfectly quiet and unwilling to move when left alone; and that this virtue supposed by him to proceed in every direction out of the Sun, swept them round, just as the sails of a windmill would carry round anything which became entangled in them. In other parts of his works Kepler mentions having speculated on a real attractive force in the centre; but as he knew that the planets are not always at the same distance from the Sun, and conceived erroneously, that to remove them from their least to their greatest distance a repulsive force must be supposed alternating with an attractive one, he laid aside this notion as improbable. In a note he acknowledges that when he wrote the passage just quoted, imbued as he then was with Scaliger's notions on moving intelligences, he literally believed "that each planet was moved by a living spirit, but afterwards came to look on the moving cause as a corporeal though immaterial substance, something in the nature of light which is observed to diminish similarly at increased distances." He then proceeds as follows in the original text.

"Let us then assume, as is very probable, that motion is dispensed by the sun in the same manner as light. The proportion in which light emanating from a centre is diminished, is taught by optical writers: for there is the same quantity of light, or of the solar rays, in the small circles as in the large; and therefore, as it is more condensed in the former, more attenuated in the latter, a measure of the attenuation may be derived from the proportion of the circles themselves, both in the case of light and of the moving virtue. Therefore, by how much the orbit of Venus is greater than that of Mercury, in the same proportion will the motion of the latter be stronger, or more hurried, or more swift, or more powerful, or by whatever other word you like to express the fact, than that of the former. But a larger orbit would require a proportionably longer time of revolution, even though the moving force were the same. Hence it follows that the one cause of a greater distance of the planet from the Sun, produces a double effect in increasing the period, and conversely the increase of the periods will be double the difference of the distances. Therefore, half the increment added to the shorter period ought to give the true proportion of the distances, so that the sum should represent the distance of the superior planet, on the same scale on which the shorter period represents the distance of the inferior one. For instance, the period of Mercury is nearly 88 days; that of Venus is 224⅔, the difference is 136⅔: half of this is 68⅓, which, added to 88, gives 156⅓. The mean distance of Venus ought, therefore, to be, in proportion to that of Mercury, as 156⅓ to 88. If this be done with all the planets, we get the following results, taking successively, as before, the distance of each planet at 1000.

The distance in parts of which the distance of the next superior planet contains 1000, is at		But according to Copernicus they are respectively
♃	574	572
♂	274	290
⊕	694	658
♀	762	719
☿	563	500

As you see, we have now got nearer the truth."

Finding that this theory of the rate of diminution would not bring him quite close to the result he desired to find, Kepler immediately imagined another. This latter occasioned him a great deal of perplexity, and affords another of the frequently recurring instances of the waste of time and ingenuity occasioned by his impetuous and precipitate temperament. Assuming the distance of any planet, as for instance of Mars, to be the unit of space, and the virtue at that distance to be the unit of force, he supposed that as many particles as the virtue at the Earth gained upon that of Mars, so many particles of distance did the Earth lose. He endeavoured to determine the respective positions of the planets upon this theory, by the rules of false position, but was much astonished at finding the same exactly as on his former hypothesis. The fact was, as he himself discovered, although not until after several years, that he had become confused in his calculation; and when half through the process, had retraced his steps so as of course to arrive again at the numbers from which he started, and which he had taken from his former results. This was the real secret of the identity of the two methods; and if, when he had taken the distance of Mars at 1000, instead of assuming the distance of the earth at 694, as he did, he had taken any other number, and operated upon it in the same manner, he would have had the same reason for relying on the accuracy of his supposition. As it was, the result utterly confounded him; and he was obliged to leave it with the remark, that " the two theories are thus proved to be the same in fact, and only different in form; although how that can possibly be, I have never to this day been able to understand."—His perplexity was very reasonable; they are by no means the same; it was only his method of juggling with the figures which seemed to connect them.

Notwithstanding all its faults, the genius and unwearied perseverance displayed by Kepler in this book, immediately ranked him among astronomers of the first class; and he received the most flattering encomiums from many of the most celebrated; among others, from Galileo and Tycho Brahe, whose opinion he invited upon his performance. Galileo contented himself with praising in general terms the ingenuity and good faith which appeared so conspicuously in it. Tycho Brahe entered into a more detailed criticism of the work, and, as Kepler shrewdly remarked, showed how highly he thought of it by advising him to try to adapt something of the same kind to the Tychonic system. Kepler also sent a copy of his book to the imperial astronomer, Raimar, with a complimentary letter, in which he exalted him above all other astronomers of the age. Raimar had surreptitiously acquired a notion of Tycho Brahe's theory, and published it as his own; and Tycho, in his letter, complained of Kepler's extravagant flattery. This drew a long apologetical reply from Kepler, in which he attributed the admiration he had expressed of Raimar to his own want of information at that time, having since met with many things in Euclid and Regiomontanus, which he then believed original in Raimar. With this explanation, Tycho professed himself perfectly satisfied.

CHAPTER II.

Kepler's Marriage—He joins Tycho Brahe at Prague—Is appointed Imperial Mathematician—Treatise on the New Star.

THE publication of this extraordinary book, early as it occurs in the history of Kepler's life, was yet preceded by his marriage. He had contemplated this step so early as 1592; but that suit having been broken off, he paid his addresses, in 1596, to Barbara Muller von Muhleckh. This lady was already a widow for the second time, although two years younger than Kepler himself. On occasion of this alliance he was required to prove the nobility of his family, and the delay consequent upon the inquiry postponed the marriage till the following year. He soon became involved in difficulties in consequence of this inconsiderate engagement: his wife's fortune was less than he had been led to expect, and he became embroiled on that account with her relations. Still more serious inconvenience resulted to him from the troubled state in which the province of Styria was at that time, arising out of the disputes in Bohemia and the two great religious parties into which the empire was now divided, the one headed by Rodolph, the feeble minded emperor,—the other by Matthias, his ambitious and enterprising brother.

In the year following his marriage, he thought it prudent, on account of some opinions he had unadvisedly promulgated, (of what nature does not very distinctly appear,) to withdraw himself from Gratz into Hungary. Thence he transmitted several short treatises to his friend Zehentmaier, at Tubingen—" On the Magnet," " On the Cause of the Obliquity of the Ecliptic," and " On the Divine Wisdom, as shown in the Creation." Little is known of these works beyond the notice taken of them in Zehentmaier's answers. Kepler has himself told us, that his magnetic philosophy was built upon the investigations of Gilbert, of whom he always justly spoke with the greatest respect.

About the same time a more violent persecution had driven Tycho Brahe from his observatory of Uraniburg, in the little island of Hueen, at the entrance of the Baltic. This had been bestowed on him by the munificence of Frederick I. of Denmark, who liberally furnished him with every means of prosecuting his astronomical observations. After Frederick's death, Tycho found himself unable to withstand the party which had constantly opposed him, and was forced, at a great loss and much inconvenience, to quit his favourite island. On the invitation of the emperor, Rudolph II., he then betook himself, after a short stay at Hamburg, to the castle of Benach, near Prague, which was assigned to him with an annual pension of three thousand florins, a truly munificent provision in those times and that country.

Kepler had been eager to see Tycho Brahe since, the latter had intimated that his observations had led him to a more accurate determination of the excentricities of the orbits of the planets. By help of this, Kepler hoped that his theory might be made to accord more nearly with the truth; and on learning that Tycho was in Bohemia, he immediately set out to visit him, and arrived at Prague in January, 1600. From thence he wrote a second letter to Tycho, not having received the answer to his former apology, again excusing himself for the part he had appeared to take with Raimar against him. Tycho replied immediately in the kindest manner, and begged he would repair to him directly: —" Come not as a stranger, but as a very welcome friend; come and share in my observations with such instruments as I have with me, and as a dearly beloved associate." During his stay of three or four months at Benach, it was settled that Tycho should apply to the emperor, to procure him the situation of assistant in the observatory. Kepler then returned to Gratz, having previously received an intimation, that he might do so in 'safety. The plan, as it had been arranged between them was, that a letter should be procured from the emperor to the states of Styria, requesting that Kepler might join Tycho Brahe for two years, and retain his salary during that time: a hundred florins were to be added annually by the emperor, on account of the greater dearness of living at Prague. But before everything was concluded, Kepler finally threw up his situation at Gratz, in consequence of new dissensions. Fearing that this would utterly put an end to his hopes of connecting himself with Tycho, he determined to revive his claims on the patronage of the Duke of Wirtemberg. With this view he entered into correspondence with Mästlin and some of his other friends at Tubingen, intending to prosecute his medical studies, and offer himself for the professorship of medicine in that university. He was dissuaded from this scheme by the pressing instances of Tycho, who undertook to exert himself in procuring a permanent settlement for him from the emperor, and assured him, 'even if that attempt should fail, that the language he had used when formerly inviting him to visit him at Hamburg, should not be forgotten. In consequence of this encouragement," Kepler abandoned his former scheme, and travelled again with his wife to Prague. He was detained a long time on the road by violent illness, and his money became entirely exhausted. On this he wrote complainingly to Tycho, that he was unable without assistance to travel even the short distance which still separated them, far less to await much longer the fulfilment of the promises held out to him.

By his subsequent admissions, it appears that for a considerable time he lived entirely on Tycho's bounty; and by way of return, he wrote an essay against Raimar, and against a Scotchman named Liddell, professor at Rostoch and Helmstadt, who, like Raimar, had appropriated to himself the credit of the Tychonic system. Kepler never adopted this theory, and indeed, as the question merely regarded 'priority of invention, there could be no occasion, in the discussion, for an examination of its principles.

This was followed by a transaction, not much to Kepler's credit, who in the course of the following year, and during a second absence from Prague, fancied that he had some reason to complain of Tycho's behaviour, and wrote him a violent letter, filled with reproaches and insults. Tycho appears to have behaved in this affair with great moderation: professing to be himself occupied with the marriage of his daughter, he gave the care of replying to Kepler's charges, to Ericksen, one of his assistants, who, in a very kind and temperate letter, pointed out to him the ingratitude of his behaviour, and the groundlessness of his dissatisfaction. His principal complaint seems to have been, that Tycho had not sufficiently supplied his wife with money during his absence. Ericksen's letter produced an immediate and entire change in Kepler's temper, and it is only from the humble recantation which he instantaneously offered that we learn the extent of his previous violence. ." Most noble Tycho," these are the words of his letter, " how shall I enumerate or rightly estimate your benefits conferred on me! For two months you have liberally and gratuitously maintained me, and my whole family; you have provided for all my wishes; you have done me every possible kindness; you have communicated to me everything you hold most dear; no one, by word or deed, has intentionally injured me in anything: in short, not to your children, your wife, or yourself have you shown more indulgence than to me. This being so, as I am anxious to put upon record, I cannot reflect without consternation that I should have been so given up by God to my own intemperance, as to shut my eyes on all these benefits; that, instead of modest and respectful gratitude, I should indulge for three weeks in continual moroseness towards all your family, in headlong passion, and the utmost insolence towards yourself, who possess so many claims on my veneration from your noble family, your extraordinary learning, and distinguished reputation. Whatever I have said or written against the person, the fame, the honour, and the learning of your excellency; or whatever, in any other way, I have injuriously spoken or written, (if they admit no other more favourable interpretation,) as to my grief I have spoken and written many things, and more than I can remember; all and everything I recant, and freely and honestly declare and profess to be groundless, false, and incapable of proof." Hoffmann, the president of the states of Styria, who had taken Kepler to Prague on his first visit, exerted himself to perfect the reconciliation, and this hasty quarrel was entirely passed over.

On Kepler's return to Prague, in September, 1601, he was presented to the Emperor by Tycho, and honoured with the title of Imperial Mathematician, on condition of assisting Tycho in his calculations. Kepler desired nothing more than this, condition, since Tycho was at that time probably the only person in the world who possessed observations sufficient for the reform which he now began to meditate in the theory of astronomy. Rudolph appears to have valued both Tycho Brahe and Kepler as astrologers rather than astronomers; but although unable to appreciate rightly the importance of the task they undertook, of compiling a new set of astronomical tables founded upon Tycho's observations, yet his vanity was flattered with the prospect of his name being connected with such a work, and he made liberal promises to defray the expense of the new Rudolphine Tables. Tycho's principal assistant at this time was Longomontanus, who altered his name to this form, according to the prevalent fashion of giving to every name a Latin termination. Lomborg or Longbierg was the name, not of his family, but of the village in Denmark, where he was born, just as Müller was seldom called by any other name than Regiomontanus, from his native town Königsberg, as George Joachim Rheticus was so surnamed from Rhetia, the country of the Grisons, and as Kepler himself was sometimes called Leonmontanus, from Leonberg, where he passed his infancy. It was agreed between Longomontanus and Kepler, that in discussing Tycho's observations, the former should apply himself especially to the Moon, and the latter to Mars, on which planet, owing to its favourable position, Tycho was then particularly engaged. The nature of these labours will be explained when we come to speak of the celebrated book " On the Motions of Mars."

This arrangement was disturbed by the return of Longomontanus into Denmark, where he had been offered an astronomical professorship, and still more by, the sudden death of Tycho Brahe himself in the following October. Kepler attended him during his illness, and after his death undertook to arrange some of his writings. But, in consequence of a misunderstanding between him and Tycho's family, the manuscripts were taken out of his hands; and when, soon afterwards, the book appeared, Kepler complained heavily that they had published, without his consent or knowledge, the notes and interlineations added by him for his own private guidance whilst preparing it for publication.

On Tycho's death, Kepler succeeded him as principal mathematician to the emperor; but although he was thus nominally provided with a liberal salary, it was almost always in arrear. The pecuniary embarrassments in which he constantly found himself involved, drove him to the resource of gaining a livelihood by casting nativities. His peculiar temperament rendered him not averse from such speculations; and he enjoyed considerable reputation in this line, and received ample remuneration for his predictions. But although he did not scruple, when consulted, to avail himself in this manner of the credulity of his contemporaries, he passed over few occasions in his works of protesting against the futility of this particular genethliac astrology. His own astrological creed was in a different strain, more singular, but not less extravagant. We shall defer entering into any details concerning it, till we come to treat of his book on Harmonics, in which he has collected and recapitulated the substance of his scattered opinions on this strange subject.

His next works deserving notice are those published on occasion of the new star which shone out with great splendour in 1604, in the constellation Cassiopeia *. Immediately on its appearance, Kepler wrote a short account of it in German, marked with all the oddity which characterises most of his productions. We shall see enough of his astronomical calculations when we come to his book on Mars; the following passage will probably be found more amusing.

After comparing this star with that of 1572, and mentioning that many persons who had seen it maintained this to be the brighter of the two, since it was nearly twice the size of its nearest neighbour, Jupiter, he proceeds as follows:—
" Yonder one chose for its appearance a time no way remarkable, and came into the world quite unexpectedly, like an enemy storming a town, and breaking into the market-place before the citizens are aware of his approach; but ours has come exactly in the year of which astrologers have written so much about the fiery trigon that happens in it †; just in the month in which (according to Cyprian) Mars comes up to a very perfect conjunction with the other two superior planets; just in the day when Mars has joined Jupiter, and just in the place where this conjunction has taken place. Therefore the apparition of this star is not like a secret hostile irruption, as was that one of 1572, but the spectacle of a public triumph, or the entry of a mighty potentate; when the couriers ride in some time before, to prepare his lodgings, and the crowd of young urchins begin to think the time over-long to wait: then roll in, one after another, the ammunition, and money, and baggage waggons, and presently the trampling of horse, and the rush of people from every side to the streets and windows; and when the crowd have gazed with their jaws all agape at the troops of knights; then at last, the trumpeters, and archers, and lackeys, so distinguish the person of the monarch, that there is no occasion to point him out, but every one cries out of his own accord—' Here we have him!'—What it may portend is hard to determine, and thus much only is certain, that it comes to tell mankind either nothing at all, or high and weighty news, quite beyond human sense and understanding. It will have an important influence on political and social relations; not indeed by its own nature, but, as it were, accidentally through the disposition of mankind. First, it portends to the booksellers great disturbances, and tolerable gains; for almost every *Theologus, Philosophicus, Medicus,* and *Mathematicus*, or whoever else, having no laborious occupation intrusted to him, seeks his pleasure *in studiis*, will make particular remarks upon it, and will wish to bring these remarks to the light. Just so will others, learned and unlearned, wish to know its meaning, and they will buy the authors who profess to tell them. I mention these things merely by way of example, because, although thus much can be easily predicted without great skill, yet may it happen just as easily, and in the same manner, that the vulgar, or whoever else is of easy faith, or it may be, crazy, may wish to exalt himself into a great prophet; or it may even happen that some powerful lord, who has good foundation and beginning of great dignities, will be cheered on by this phenomenon to venture on some new scheme, just as if God had set up this star in the darkness merely to enlighten them."

It would hardly be supposed, from the tenor of this last passage, that the writer of it was not a determined enemy to astrological predictions of every description. In 1602 he had published a disputation, not now easily met with, "On the Principles of Astrology," in which it seems that he treated the professed astrologers with great severity. The essence of this book is probably contained in the second treatise on the new star, which he published in 1606*. In this volume he inveighs repeatedly against the vanity and worthlessness of ordinary astrology, declaring at the same time, that the professors of that art know that this judgment is pronounced by one well acquainted with its principles. "For if the vulgar are to pronounce who is the best astrologer, my reputation is known to be of the highest order; if they

---

* See Life of Galileo, p. 16.
† The fiery trigon occurs about once in every 800 years, when Saturn, Jupiter, and Mars are in the three fiery signs, Aries, Leo, and Sagittarius.

* The copy of this work in the British Museum is Kepler's presentation copy to our James I. On the blank leaf, opposite the title-page, is the following inscription, apparently in the author's handwriting:—" Regi philosophanti, philosophus serviens, Platoni Diogenes, Britannias tenenti, Pragæ stipem mendicans ab Alexandro, e dolio conductitio, hoc suum philosophema misit et commendavit."

prefer the judgment of the learned, they are already condemned. Whether they stand with me in the eyes of the populace, or I fall with them before the learned, in both cases I am in their ranks; I am on a level with them; I cannot be renounced."

The theory which Kepler proposed to substitute is intimated shortly in the following passage: " I maintain that the colours and aspects, and conjunctions of the planets, are impressed on the natures or faculties of sublunary things, and when they occur, that these are excited as well in forming as in moving the body over whose motion they preside. Now let no one conceive a prejudice that I am anxiously seeking to mend the deplorable and hopeless cause of astrology, by far-fetched subtilties and miserable quibbling. I do not value it sufficiently, nor have I ever shunned having astrologers for my enemies. But a most unfailing experience (as far as can be hoped in natural phenomena) of the excitement of sublunary natures by the conjunctions and aspects of the planets, has instructed and compelled my unwilling belief."

After exhausting other topics suggested by this new star, he examines the different opinions on the cause of its appearance. Among others he mentions the Epicurean notion, that it was a fortuitous concourse of atoms, whose appearance in this form was merely one of the infinite number of ways in which, since the beginning of time, they have been combined. Having descanted for some time on this opinion, and declared himself altogether hostile to it, Kepler proceeds as follows:—" When I was a youth, with plenty of idle time on my hands, I was much taken with the vanity, of which some grown men are not ashamed, of making anagrams, by transposing the letters of my name, written in Greek, so as to make another sentence: out of Ἰωάννης Κεπλῆρος I made Σειρήνων κάπηλος*; in Latin, out of *Joannes Keplerus* came *Serpens in akuleo*†. But not being satisfied with the meaning of these words, and being unable to make another, I trusted the thing to chance, and taking out of a pack of playing cards as many as there were letters in the name, I wrote one upon each, and then began to shuffle them, and at each shuffle to read them in the order they came, to see if any meaning came of it. Now, may all the Epicurean gods and goddesses confound this same chance, which, although I spent a good deal of time over it, never showed me anything like sense even from a distance *. So I gave up my cards to the Epicurean eternity, to be carried away into infinity, and, it is said, they are still flying about there, in the utmost confusion among the atoms, and have never yet come to any meaning. I will tell these disputants, my opponents, not my own opinion, but my wife's. Yesterday, when weary with writing, and my mind quite dusty with considering these atoms, I was called to supper, and a salad I had asked for was set before me. It seems then, said I aloud, that if pewter dishes, leaves of lettuce, grains of salt, drops of water, vinegar, and oil, and slices of egg, had been flying about in the air from all eternity, it might at last happen by chance that there would come a salad. Yes, says my wife, but not so nice and well dressed as this of mine is."

CHAPTER III.

Kepler publishes his Supplement to Vitellion—Theory of Refraction.

DURING several years Kepler remained, as he himself forcibly expressed it, begging his bread from the emperor at Prague, and the splendour of his nominal income served only to increase his irritation, at the real neglect under which he nevertheless persevered in his labours. His family was increasing, and he had little wherewith to support them beyond the uncertain proceeds of his writings and nativities. His salary was charged partly on the states of Silesia, partly on the imperial treasury; but it was in vain that repeated orders were procured for the payment of the arrears due to him. The resources of the empire were drained by the constant demands of an engrossing war, and Kepler had not sufficient influence to enforce his claims against those who thought even the smallest sum bestowed upon him ill spent, in fostering profitless speculations. In consequence of this niggardliness, Kepler was forced to postpone the publication of the Rudolphine Tables, which he was engaged in constructing from his own and Tycho Brahe's observations, and applied himself to other works of a less costly description. Among these may be men-

---

* The tapster of the Sirens.
† A serpent in his sting.

* In one of his anonymous writings Kepler has anagrammatized his name, *Joannes Keplerus*, in a variety of other forms, probably selected from the luckiest of his shuffles :—" *Kleopas Herennius, Helenor Kapuensis, Raspinus Enkeleo, Kanones Pueriles.*"

tioned a "Treatise on Comets," written on occasion of one which appeared in 1607; in this he suggests that they are planets moving in straight lines. The book published in 1604, which he entitles "A Supplement to Vitellion," may be considered as containing the first reasonable and consistent theory of optics, especially in that branch of it usually termed dioptrics, which relates to the theory of vision through transparent substances. In it was first explained the true use of the different parts of the eye, to the knowledge of which Baptista Porta had already approached very nearly, though he stopped short of the accurate truth. Kepler remarked the identity of the mechanism in the eye with that beautiful invention of Porta's, the camera obscura; showing, that the light which falls from external objects on the eye is refracted through a transparent substance, called, from its form and composition, the crystalline lens, and makes a picture on the fine net-work of nerves, called the retina, which lies at the back of the eye. The manner in which the existence of this coloured picture on the retina causes to the individual the sensation of sight, belongs to a theory not purely physical; and beyond this point Kepler did not attempt to go.

The direction into which rays of light (as they are usually called) are bent or refracted in passing through the air and other transparent substances or mediums, is discussed in this treatise at great length. Tycho Brahe had been the first astronomer who recognized the necessity of making some allowance on this account in the observed heights of the stars. A long controversy arose on this subject between Tycho Brahe and Rothman, the astronomer at Hesse Cassel, a man of unquestionable talent, but of odd and eccentric habits. Neither was altogether in the right; although Tycho had the advantage in the argument. He failed however to establish the true law of refraction; and Kepler has devoted a chapter to an examination of the same question. It is marked by precisely the same qualities as those appearing so conspicuously in his astronomical writings:—great ingenuity; wonderful perseverance; bad philosophy. That this may not be taken solely upon assertion, some samples of it are subjoined. The writings of the authors of this period are little read or known at the present day; and it is only by copious extracts that any accurate notion can be formed of the nature and value of their labours.

The following tedious specimen of Kepler's mode of examining physical phenomena is advisedly selected to contrast with his astronomical researches: though the luck and consequently the fame that attended his divination were widely different on the two occasions, the method pursued was the same. After commenting on the points of difference between Rothman and Tycho Brahe, Kepler proceeds to enumerate his own endeavours to discover the law of refraction.

"I did not leave untried whether, by assuming a horizontal refraction according to the density of the medium, the rest would correspond with the sines of the distances from the vertical direction, but calculation proved that it, was not so: and indeed there was no occasion to have tried it, for thus the refractions would increase according to the same law in all mediums, which is contradicted by experiment.

"The same kind of objection may be brought against the cause of refraction alleged by Alhazen and Vitellion. They say that the light seeks to be compensated for the loss sustained at the oblique impact; so that in proportion as it is enfeebled by striking against the denser medium, in the same degree does it restore its energy by approaching the perpendicular; that it may strike the bottom of the denser medium with greater force; for those impacts are most forcible which are direct. And they add some subtle notions, I know not what, how the motion of obliquely incident light is compounded of a motion perpendicular and a motion parallel to the dense surface; and that this compound motion is not destroyed, but only retarded by meeting the denser medium.

"I tried another way of measuring the refraction; which should include the density of the medium and the incidence:

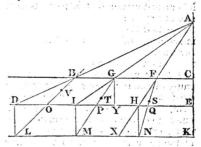

for, since a denser medium is the cause of refraction, it seems to be the same thing as if we were to prolong the depth of the medium in which the rays are refracted into as much space as would be filled by the denser medium under the force of the rarer one. Let A be the place of the light, B C the surface of the denser medium, D E its bottom. Let A B, A G, A F be rays falling obliquely, which would arrive at D, I, H, if the medium were uniform. But because it is denser; suppose the bottom to be depressed to K L, determined by this that there is as much of the denser matter contained in the space DC as of the rarer in LC: and thus, on the sinking of the whole bottom DE, the points D, I, H, E will descend vertically to L, M, N, K. Join the points. B L, G M, F N, cutting D E in O, P, Q ; the refracted rays will be A B O, A G P, A F Q."—"This method is refuted by experiment ; it gives the refractions near the perpendicular A C too great in respect of those near the horizon. Whoever has leisure may verify this, either by calculation or compasses. It may be added that the reasoning itself is not very sure-footed, and, whilst seeking to measure other things, scarcely takes in and comprehends itself." This reflection must not be mistaken for the dawn of suspicion that his examination of philosophical questions began not altogether at the right end: it is merely an acknowledgment that he had not yet contrived a theory with which he was quite satisfied before it was disproved by experiment. After some experience of Kepler's miraculous good fortune in seizing truths across the wildest and most absurd theories, it is not easy to keep clear of the opposite feeling of surprise whenever any of his extravagancies fail to discover to him some beautiful law of nature. But we must follow him as he plunges deeper in this unsuccessful inquiry ; and the reader must remember, in order fully to appreciate this method of philosophizing, that it is almost certain that Kepler laboured upon every one of the gratuitous suppositions that he makes, until positive experiment satisfied him of their incorrectness.

"I go on to other methods. Since density is clearly connected with the cause of the refractions, and refraction itself seems a kind of compression of light, as it were, towards the perpendicular, it occurred to me to examine whether there was the same proportion between the mediums in respect of density and the parts of the bottom illuminated by the light, when let into a vessel, first empty, and afterwards filled with water.

This mode branches out into many : for the proportion may be imagined, either in the straight lines, as if one should say that the line E Q, illuminated by refraction, is to EH illuminated directly, as the density of the one medium is to that of the other—Or another may suppose the proportion to be between F C and F H—Or it may be conceived to exist among surfaces, or so that some power of E Q should be to some power of E H in this proportion, or the circles or similar figures described on them. In this manner the proportion of E Q to E P would be double that of E H to EI—Or the proportion may be conceived existing among the solidities of the pyramidal frustums F H E C, F Q E C—Or, since the proportion of the mediums involves a threefold consideration, since they have density in length, breadth, and thickness, I proceeded also to examine the cubic proportions among the lines E Q, E H.

"I also considered other lines. From any of the points of refraction as G, let a perpendicular GY be dropped upon the bottom. It may become a question whether possibly the triangle I G Y, that is, the base I Y, is divided by the refracted ray G P, in the proportion of the densities of the mediums.

"I have put all these methods here together, because the same remark disproves them all. For, in whatever manner, whether as line, plane, or pyramid, E F observes a given proportion to E P, or the abbreviated line Y I to Y P, namely, the proportion of the mediums, it is sure that E I; the tangent of the distance of the point A from the vertex, will become infinite, and will, therefore make E P or Y P, also infinite. Therefore, I G P, the angle of refraction, will be entirely lost ; and, as it approaches the horizon, will gradually become less and less, which is contrary to experiment.

. "I tried again whether the images are equally removed from their points of refraction, and whether the ratio of the densities measures the least distance. For instance, supposing E to be the image, C the surface of the water, K the bottom; and C E to C K in the proportion of the densities of the mediums. Now, let F, G, B, be three other points of refraction and images at S, T, V, and let C E be equal to F S, G T, and B V. But according to this rule an image E would still be somewhat raised in the perpendicular A K, which is contrary to experiment, not to mention other contradictions. Thirdly, whether the proportion of the mediums holds between FH and FX, supposing H to be the place of the image? Not at all. For so, CE would be in the same proportion to CK, so that the height of the image would always be the same, which we have just refuted. Fourthly, whether the raising of the image at E is to the raising at H, as CE to FH? Not in the least; for so the images either would never begin to be raised, or, having once begun, would at last be infinitely raised, because FH at last becomes infinite. Fifthly, whether the images rise in proportion to the sines of the inclinations? Not at all; for so the proportion of ascent would be the same in all mediums. Sixthly, are then the images raised at first, and in perpendicular radiation, according to the proportion of the mediums, and do they subsequently rise more and more according to the sines of the inclinations? For so the proportion would be compound, and would become different in different mediums. There is nothing in it: for the calculation disagreed with experiment. And generally it is in vain to have regard to the image or the place of the image, for that very reason, that it is imaginary. For there is no connexion between the density of the medium or any real [quality or refraction of the light, and an accident of vision, by an error of which the image happens.

"Up to this point, therefore, I had followed a nearly blind mode of inquiry, and had trusted to good fortune; but now I opened the other eye, and hit upon a sure method, for I pondered the fact, that the image of a thing seen under water approaches closely to the true ratio of the refraction, and almost measures it; that it is low if the thing is viewed directly from above; that by degrees it rises as the eye passes towards the horizon of the water. Yet, on the other hand, the reason alleged above, proves that the measure is not to be sought in the image, because the image is not a thing actually existing, but arises from a deception of vision which is purely accidental. By a comparison of these conflicting arguments, it occurred to me at length, to seek the causes themselves of the existence of the image under water, and in these causes the measure of the refractions. This opinion was strengthened in me by seeing that opticians had not rightly pointed out the cause of the image which appears both in mirrors and in water. And this was the origin of that labour which I undertook in the third chapter. Nor, indeed, was that labour trifling, whilst hunting down false opinions of all sorts among the principles, in a matter rendered so intricate by the false traditions of optical writers; whilst striking out half a dozen different paths, and beginning anew the whole business. How often did it happen that a rash confidence made me look upon that which I sought with such ardour, as at length discovered!

"At length I cut this worse than Gordian knot of catoptrics by analogy alone, by considering what happens in mirrors, and what must happen analogically in water. In mirrors, the image appears at a distance from the real place of the object, not being itself material, but produced solely by reflection at the polished surface. Whence it followed in water also, that the images rise and approach the surface, not according to the law of the greater or less density in the water, as the view is less or more oblique, but solely because of the refraction of the ray of light passing from the object to the eye. On which assumption, it is plain that every attempt I had hitherto made to measure refractions by the image, and its elevation, must fall to the ground. And this became more evident when I discovered the true reason why the image is in the same perpendicular line with the object both in mirrors and in dense mediums. When I had succeeded thus far by analogy in this most difficult investigation, as to the place of the image, I began to follow out the analogy further, led on by the strong desire of measuring refraction. For I wished to get hold of some measure of some sort, no matter how blindly, having no fear but that so soon as the measure should be accurately known, the cause would plainly appear. I went to work as follows. In convex mirrors the image is diminished, and just so in rarer mediums; in denser mediums it is magnified, as in concave mirrors. In convex mirrors the central parts of the image approach, and recede in concave farther than towards the circumference; the same thing happens in different mediums, so that in water the bottom appears depressed, and the surrounding parts elevated. Hence it appears that a denser medium corresponds with a concave reflecting surface, and a rarer one with a convex one: it was clear, at the same time, that the plane surface of the water affects a property of curvature. I was, therefore, to excogitate causes consistent with its having this effect of curvature, and to see if a reason could be given, why the parts of the water surrounding the incident perpendicular should represent a greater density than the parts just under the perpendicular. And so the thing came round again to my former attempts, which being refuted by reason and experiment, I was forced to abandon the search after a cause. I then proceeded to measurements."

Kepler then endeavoured to connect his measurements of different quantities of refraction with the conic sections, and was tolerably well pleased with some of his results. They were however not entirely satisfactory, on which he breaks off with the following sentence: "Now, reader, you and I have been detained sufficiently long whilst I have been attempting to collect into one faggot the measure of different refractions: I acknowledge that the cause cannot be connected with this mode of measurement: for what is there in common between refractions made at the plane surfaces of transparent mediums, and mixtilinear conic sections? Wherefore, *quod Deus bene vortat*, we will now have had enough of the causes of this measure; and although, even now, we are perhaps erring something from the truth, yet it is better, by working on, to show our industry, than our laziness by neglect."

Notwithstanding the great length of this extract, we must add the concluding paragraph of the Chapter, directed, as we are told in the margin, against the "Tychonomasticks:"—

" I know how many blind men at this day dispute about colours, and how they long for some one to give some assistance by argument to their rash insults of Tycho, and attacks upon this whole matter of refractions; who, if they had kept to themselves their puerile errors and naked ignorance, might have escaped censure; for that may happen to many great men. But since they venture forth publicly, and with thick books and sounding titles, lay baits for the applause of the unwary, (for now-a-days there is more danger from the abundance of bad books, than heretofore from the lack of good ones,) therefore let them know that a time is set for them publicly to amend their own errors. If they longer delay doing this, it shall be open, either to me or any other, to do to these unhappy meddlers in geometry as they have taken upon themselves to do with respect to men of the highest reputation. And although this labour will be despicable, from the vile nature of the follies against which it will be directed, yet so much more necessary than that which they have undertaken against others, as he is a greater public nuisance, who endeavours to slander good and necessary inventions, than he who fancies he has found what is impossible to discover. Meanwhile, let them cease to plume themselves on the silence which is another word for their own obscurity."

Although Kepler failed, as we have seen, to detect the true law of refraction, (which was discovered some years later by Willibrord Snell, a Flemish mathematician,) there are many things well deserving notice in his investigations. He remarked, that the quantity of refraction would alter, if the height of the atmosphere should vary; and also, that it would be different at different temperatures. Both these sources of variation are now constantly taken into account, the barometer and thermometer giving exact indications of these changes. There is also a very curious passage in one of his letters to Bregger, written in 1605, on the subject of the colours in the rainbow. It is in these words:—
" Since every one sees a different rainbow, it is possible that some one may see a rainbow in the very place of my sight. In this case, the medium is coloured at the place of my vision, to which the solar ray comes to me through water, rain, or aqueous vapours. For the rainbow is seen when the sun is shining between rain, that is to say, when the sun also is visible. Why then do I not see the sun green, yellow, red, and blue; if vision takes place according to the mode of illumination? I will say something for you to attack or examine. The sun's rays are not coloured, except with a definite quantity of refraction. Whether you are in the optical chamber, or standing opposite glass globes, or walking in the morning dew, everywhere it is obvious that a certain and definite angle is observed, under which, when seen in dew, in glass, in water, the sun's splendour appears coloured, and under no other angle. There is no colouring by mere reflexion, without the refraction of a denser medium." How closely does Kepler appear, in this passage, to approach the discovery which forms not the least part of Newton's fame!

We also find in this work a defence of the opinion that the planets are lumi nous of themselves; on the ground that the inferior planets would, on the contrary supposition, display phases like those of the moon when passing between us and the sun. The use of the telescope was not then known; and, when some years later the form of the disk of the planets was more clearly defined with their assistance, Kepler had the satisfaction of finding his assertions verified by the discoveries of Galileo, that these changes do actually take place. In another of his speculations, connected with the same subject, he was less fortunate. In 1607 a black spot appeared on the face of sun, such as may almost always be seen with the assistance of the telescope, although they are seldom large enough to be visible to the unassisted eye. Kepler saw it for a short time, and mistook it for the planet Mercury, and with his usual precipitancy hastened to publish an account of his observation of this rare phenomenon. A few years later, Galileo discovered with his glasses, a great number of similar spots; and Kepler immediately retracted the opinion announced in his treatise, and acknowledged his belief that previous accounts of the same occurrence which he had seen in old authors, and which he had found great difficulty in reconciling with his more accurate knowledge of the motions of Mercury, were to be referred to a like mistake. On this occasion of the invention of the telescope, Kepler's candour and real love of truth appeared in a most favourable light. Disregarding entirely the disagreeable necessity, in consequence of the discoveries of this new instrument, of retracting several opinions which he had maintained with considerable warmth, he ranged himself at once on the side of Galileo, in opposition to the bitter and determined hostility evinced by most of those whose theories were endangered by the new views thus offered of the heavens. Kepler's quarrel with his pupil, Horky, on this account, has been mentioned in the "Life of Galileo;" and this is only a selected instance from the numerous occasions on which he espoused the same unpopular side of the argument. He published a dissertation to accompany Galileo's "Intelligencer of the Stars," in which he warmly expressed his admiration of that illustrious inquirer into nature. His conduct in this respect was the more remarkable, as some of his most intimate friends had taken a very opposite view of Galileo's merit, and seem to have laboured much to disturb their mutual regard; Mästlin especially, Kepler's early instructor, seldom mentioned to him the name of Galileo, without some contemptuous expression of dislike. These statements have rather disturbed the chronological order of the account of Kepler's works. We now return to the year 1609, in which he published his great and extraordinary book, "On the Motions of Mars;" a work which holds the intermediate place, and is in truth the connecting link, between the discoveries of Copernicus and Newton.

CHAPTER IV.
Sketch of the Astronomical Theories before Kepler.

KEPLER had begun to labour upon these commentaries from the moment when he first made Tycho's acquaintance; and it is on this work that his reputation should be made mainly to rest. It is marked in many places with his characteristic precipitancy, and indeed one of the most important discoveries announced in it (famous among astronomers by the name of the Equable Description of Areas) was blundered upon by a lucky compensation of errors; of the nature of which Kepler remained ignorant to the very last. Yet there is more of the inductive method in this than in any of his other publications; and the unwearied perseverance with which he exhausted years in hunting down his often renewed theories, till at length he seemed to arrive at the true one, almost by having previously disproved every other, excites a feeling of astonishment nearly approaching to awe. It is wonderful how he contrived to retain his vivacity and creative fancy amongst the clouds of figures which he conjured up round him; for the slightest hint or shade of probability was sufficient to plunge him into the midst of the most laborious computations. He was by no means an accurate calculator, according to the following character which he has given of himself:—"Something of these delays must be attributed to my own temper, for *non omnia possumus omnes*, and I am totally unable to observe any order; what I do suddenly, I do confusedly, and if I produce any thing well arranged, it has been done ten times over. Sometimes an error of calculation committed by hurry, delays me a great length of time. I could indeed publish an infinity of things, for though my reading is confined, my imagination is abundant, but I grow dissatisfied with such confusion: I get disgusted and out of humour, and either throw them away, or put them aside to be looked at again; or, in other words, to be written again, for that is generally the end of it. I entreat you, my friends, not to condemn me for ever to grind in the mill of mathematical calculations: allow me some time for philosophical speculations, my only delight."

He was very seldom able to afford the expense of maintaining an assistant, and was forced to go through most of the drudgery of his calculations by himself; and the most confirmed and merest arithmetician could not have toiled more doggedly than Kepler did in the work of which we are about to speak.

In order that the language of his astronomy may be understood, it is necessary to mention briefly some of the older theories. When it had been discovered that the planets did not move regularly round the earth, which was supposed to be fixed in the centre of the world, a mechanism was contrived by which it was thought that the apparent irregularity could be represented, and yet the principle of uniform motion, which was adhered to with superstitious reverence, might be preserved. This, in its simplest form, consisted in supposing the planet to move uniformly in a small circle, called an *epicycle*, the centre of which moved with an equal angular motion in the opposite direction round the earth*. The circle D d, described by D, the centre of the epicycle, was called the *deferent*. For instance, if the planet was supposed to be at A when the centre of the epicycle was at D, its

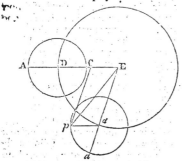

position, when the centre of the epicycle had removed to d, would be at p, found by drawing d p parallel to D A. Thus, the angle a d p, measuring the motion of the planet in its epicycle, would be equal to D E d, the angle described by the centre of the epicycle in the deferent. The angle p E d between E p, the direction in which a planet so moving would be seen from the earth, supposed to be at E, and E d, the direction in which it would have been seen had it been moving in the centre of the deferent, was called the equation of the orbit, the word equation, in the language of astronomy, signifying what must be added or taken from an irregularly varying quantity to make it vary uniformly.

As the accuracy of observations increased, minor irregularities were discovered, which were attempted to be accounted for by making a second deferent of the epicycle, and making the centre of a second epicycle revolve in the circumference of the first, and so on, or else by supposing the revolution in the epicycle not to be completed in exactly the time in which its centre is carried round the deferent. Hipparchus was the first to make a remark by which the geometrical representation of these inequalities was considerably simplified. In fact, if E C be taken equal to p d, C d will be a parallelogram, and consequently C p equal to E d, so that the machinery of the first deferent and epicycle amounts to supposing that the planet revolves uniformly in a circle round the point C, not coincident with the place of the earth. This was consequently called the excentric theory, in opposition to the former or concentric one, and was received as a great improvement. As the point d is not represented by this construction, the equation to the orbit was measured by the angle C p E, which is equal to p E d. It is not necessary to give any account of the manner in which the old astronomers determined the magnitudes and positions of these orbits, either in the concentric or excentric theory, the present object being little more than to explain the meaning of the terms it will be necessary to use in describing Kepler's investigations.

To explain the irregularities observed in the other planets, it became necessary to introduce another hypothesis, in adopting which the severity of the principle of uniform motion was somewhat relaxed. The machinery consisted partly of an excentric deferent round E, the earth, and on it an epicycle, in which the planet revolved uniformly; but the centre of the epicycle, instead of revolving uniformly round C, the centre of the deferent,

---

* By "the opposite direction" is meant, that while the motion in the circumference of one circle appeared, as viewed from its centre, to be from left to right, the other, viewed from its centre, appeared from right to left. This must be understood whenever these or similar expressions are repeated.

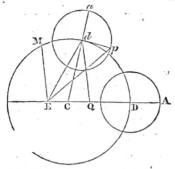

as it had hitherto been made to do, was supposed to move in its circumference with an uniform angular motion round a third point, Q; the necessary effect of which supposition was, that the linear motion of the centre of the epicycle ceased to be uniform. There were thus three points to be considered within the deferent; E, the place of the earth; C, the centre of the deferent, and sometimes called the centre of the orbit; and Q, called the centre of the equant, because, if any circle were described round Q, the planet would appear to a spectator at Q, to be moving equably in it. It was long uncertain what situation should be assigned to the centre of the equant, so as best to represent the irregularities to a spectator on the earth, until Ptolemy decided on placing it (in every case but that of Mercury, the observations on which were very doubtful) so that C, the centre of the orbit, lay just half way in the straight line, joining Q, the centre of equable motion, and E, the place of the earth. This is the famous principle, known by the name of the bisection of the excentricity.

The first equation required for the planet's motion was thus supposed to be due to the displacement of E, the earth, from Q, the centre of uniform motion, which was called the excentricity of the equant: it might be represented by the angle d E M, drawing E M parallel to Q d; for clearly M would have been the place of the centre of the epicycle at the end of a time proportional to D d, had it moved with an equable angular motion round E instead of Q. This angle d E M, or its equal E d Q, was called the equation of the centre (i. e. of the centre of the epicycle); and is clearly greater than if E Q, the excentricity of the equant, had been no greater than E C, called the excentricity of the orbit. The second equation was measured by the angle subtended at E by d, the centre of the epicycle, and p the planet's place in its circumference: it was called indifferently the equation of the orbit, or of the argument. In order to account for the apparent stations and retrogradations of the planets, it became necessary to suppose that many revolutions in the latter were completed during one of the former. The variations of latitude of the planets were exhibited by supposing not only that the planes of their deferents were oblique to the plane of the ecliptic, and that the plane of the epicycle was also oblique to that of the deferent, but that the inclination of the two latter was continually changing, although Kepler doubts whether this latter complication was admitted by Ptolemy. In the inferior planets, it was even thought necessary to give to the plane of the epicycle two oscillatory motions on axes at right angles to each other.

The astronomers at this period were much struck with a remarkable connexion between the revolutions of the superior planets in their epicycles, and the apparent motion of the sun; for when in conjunction with the sun, as seen from the earth, they were always found to be in the apogee, or point of greatest distance from the earth, of their epicycle; and when in opposition to the Sun, they were as regularly in the perigee, or point of nearest approach of the epicycle. This correspondence between two phenomena, which, according to the old astronomy, were entirely unconnected, was very perplexing, and it seems to have been one of the facts which led Copernicus to substitute the theory of the earth's motion round the sun.

As time wore on, the superstructure of excentrics and epicycles, which had been strained into representing the appearances of the heavens at a particular moment, grew out of shape, and the natural consequence of such an artificial system was, that it became next to impossible to foresee what ruin might be produced in a remote part of it by any attempt to repair the derangements and refit the parts to the changes, as they began to be remarked in any particular point. In the ninth century of our era, Ptolemy's tables were already useless, and all those that were contrived with unceasing toil to supply their place, rapidly became as unserviceable as they. Still the triumph of genius was seen in the veneration that continued to be paid to the assumptions of Ptolemy and Hipparchus; and even when the great reformer, Copernicus, appeared, he did not for a long time intend to do more than slightly modify their principles. That which he found difficult in the Ptolemaic system, was none of the inconveniences by which, since the establishment of the new system, it has become common to demonstrate the inferiority of the old one; it was the displacement of the centre of the equant from the centre of the orbit that principally indisposed him against it, and led him to endeavour to represent the appearances by some other combinations of really uniform circular motions.

There was an old system, called the Egyptian, according to which Saturn, Jupiter, Mars, and the Sun circulated round the earth, the sun carrying with it, as two moons or satellites, the other two planets, Venus and Mercury. This system had never entirely lost credit: it had been maintained in the fifth century by Martianus Capella*, and indeed it was almost sanctioned, though not formally taught, by Ptolemy himself, when he made the mean motion of the sun the same as that of the centres of the epicycles of both these planets. The remark which had also been made by the old astronomers, of the connexion between the motion of the sun and the revolutions of the superior p   in their epicycles, led him straightanetthe expectation that he might, perhaps, produce the uniformity he sought by extending the Egyptian system to these also, and this appears to have been the shape in which his reform was originally projected. It was already allowed that the centre of the orbits of all the planets was not coincident with the earth, but removed from it by the space E C. This first change merely made E C the same for all the planets, and equal to the mean distance of the earth from the sun. This system afterwards acquired great celebrity through its adoption by Tycho Brahe, who believed it originated with himself. It might perhaps have been at this period of his researches, that Copernicus was struck with the passages in the Latin and Greek authors, to which he refers as testifying the existence of an old belief in the motion of the earth round the sun. He immediately recognised how much this alteration would further his principles of uniformity, by referring all the planetary motions to one centre, and did not hesitate to embrace it. The idea of explaining the daily and principal apparent motions of the heavenly bodies by the revolution of the earth on its axis, would be the concluding change, and became almost a necessary consequence of his previous improvements, as it was manifestly at variance with his principles to give to all the planets and starry worlds a rapid daily motion round the centre of the earth, now that the latter was removed from its former supposed post in the centre of the universe, and was itself carried with an annual motion round another fixed point.

The reader would, however, form an inaccurate notion of the system of Copernicus, if he supposed that it comprised no more than the theory that each planet, including the earth among them, revolved in a simple circular orbit round the sun. Copernicus was too well acquainted with the motions of the heavenly bodies, not to be aware that such orbits would not accurately represent them; the motion he attributed to the earth round the sun, was at first merely intended to account for those which were called the second inequalities of the planets, according to which they appear one while to move forwards, then backwards, and at intermediate periods, stationary, and which thenceforward were also called the optical equations, as being merely an optical illusion. With regard to what were called the first inequalities, or physical equations, arising from a real inequality of motion, he still retained the machinery of the deferent and epicycle; and all the alteration he attempted in the orbits of the superior planets was an extension of the concentric theory to supply the place of the equant, which he considered the blot of the system. His theory for this purpose is shown in the accompanying diagram, where S represents the sun,

* Venus Mercuriusque, licet ortus occasusque quotidianos ostendunt, tamen eorum circuli terras omnino non ambiunt, sed circa solem laxiore ambitu circulantur. Denique circulorum suorum centron in sole constituunt.—De Nuptiis Philologiæ et Mercurii. Vicentiæ. 1499.

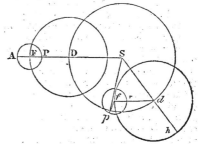

D d, the deferent or mean orbit of the

planet, on which revolves the centre of the great epicycle, whose radius, D F, was taken at ¼ of Ptolemy's excentricity of the equant; and round the circumference of this revolved, in the opposite direction, the centre of the little epicycle, whose radius, F P, was made equal to the remaining ¼ of the excentricity of the equant.

The planet P revolved in the circumference of the little epicycle, in the same direction with the centre of the great epicycle in the circumference of the deferent, but with a double angular velocity. The planet was supposed to be in the perigee of the little epicycle, when its centre was in the apogee of the greater; and whilst, for instance, D moved equably though the angle DSd, F moved through $h\ df = DSd$, and P through $rfp = 2\ DSd$.

It is easy to show that this construction gives nearly the same result as Ptolemy's; for the deferent and great epicycle have been already shown exactly equivalent to an excentric circle round S, and indeed Copernicus latterly so represented it: the effect of his construction, as given above, may therefore be reproduced in the following simpler form, in which only the smaller epicycle is retained:

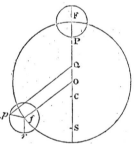

In this construction, the place of the planet is found at the end of any time proportional to Ff, by drawing fr parallel to SF, and taking $rfp = 2\ Fof$. Hence it is plain, if we take O Q, equal to F P, (already assumed equal to ¼ of Ptolemy's excentricity of the equant,) since S O is equal to ¾ of the same, that S Q is the whole of Ptolemy's excentricity of the equant; and therefore, that Q is the position of the centre of his equant. It is also plain if we join Qp, since $rfp = 2Fof$, and $oQ = fp$, that p Q is parallel to fo, and, therefore, p Q P is proportional to the time; so that the planet moves uniformly about the same point Q, as in Ptolemy's theory; and if we bisect S Q

in C, which is the position of the centre of Ptolemy's deferent, the planet will, according to Copernicus, move very nearly, though not exactly, in the same circle, whose radius is C P, as that given by the simple excentric theory.

The explanation offered by Copernicus, of the motions of the inferior planets, differed again in form from that of the others. He here introduced what was called a *hypocycle*; which, in fact, was nothing but a deferent not including the sun, round which the centre of the orbit revolved. An epicycle in addition to the hypocycle was introduced into Mercury's orbit. In this epicycle he was not supposed to revolve, but to librate, or move up and down in its diameter. Copernicus had recourse to this complication to satisfy an erroneous assertion of Ptolemy with regard to some of Mercury's inequalities. He also retained the oscillatory motions ascribed by Ptolemy to the planes of the epicycles, in order to explain the unequal latitudes observed at the same distance from the nodes, or intersections of the orbit of the planet with the ecliptic. Into this intricacy, also, he was led by placing too much confidence in Ptolemy's observations, which he was unable to satisfy by an unvarying obliquity. Other very important errors, such as his belief that the line of nodes always coincided with the line of apsides, or places of greatest and least distance from the central body, (whereas, at that time, in the case of Mars, for instance, they were nearly 90? asunder,) prevented him from accurately representing many of the celestial phenomena.

These brief details may serve to show that the adoption or rejection of the theory of Copernicus was not altogether so simple a question as sometimes it may have been considered. It is, however, not a little remarkable, while it is strongly illustrative of the spirit of the times, that these very intricacies, with which Kepler's theories have enabled us to dispense, were the only parts of the system of Copernicus that were at first received with approbation. His theory of Mercury, especially, was considered a masterpiece of subtle invention. Owing to his dread of the unfavourable judgment he anticipated on the main principles of his system, his work remained unpublished during forty years, and was at last given to the world only just in time to allow Copernicus to receive the first copy of it a few hours before his death.

CHAPTER V.

Account of the Commentaries on the motions of Mars—Discovery of the Law of the equable description of Areas, and of Elliptic Orbits.

We may now proceed to examine Kepler's innovations, but it would be doing injustice to one of the brightest points of his character, not to preface them by his own animated exhortation to his readers. "If any one be too dull to comprehend the science of astronomy, or too feeble-minded to believe in Copernicus without prejudice to his piety, my advice to such a one is, that he should quit the astronomical schools, and condemning, if he has a mind, any or all of the theories of philosophers, let him look to his own affairs, and leaving this worldly travail, let him go home and plough his fields: and as often as he lifts up to this goodly heaven those eyes with which alone he is able to see, let him pour out his heart in praises and thanksgiving to God the Creator; and let him not fear but he is offering a worship not less acceptable than his to whom God has granted to see yet more clearly with the eyes of his mind, and who both can and will praise his God for what he has so discovered."

Kepler did not by any means underrate the importance of his labours, as is sufficiently shewn by the sort of colloquial motto which he prefixed to his work. It consists in the first instance of an extract from the writings of the celebrated and unfortunate Peter Ramus. This distinguished philosopher was professor of mathematics in Paris, and in the passage in question, after calling on his contemporaries to turn their thoughts towards the establishment of a system of Astronomy unassisted by any hypothesis, he promised as an additional inducement to vacate his own chair in favour of any one who should succeed in this object. Ramus perished in the massacre of St. Bartholomew, and Kepler apostrophizes him as follows:—" It is well, Ramus, that you have forfeited your pledge, by quitting your life and professorship together: for if you still held it, I would certainly claim it as of right belonging to me on account of this work, as I could convince you even with your own logic." It was rather bold in Kepler to assert his claim to a reward held out for a theory resting on no hypothesis, by right of a work filled with hypotheses of the most startling description; but of the vast importance of this book there can be no doubt; and throughout the many wild and eccentric ideas to which we are introduced in the course of it, it is fit always to bear in mind that they form part of a work which is almost the basis of modern Astronomy."

The introduction contains a curious criticism of the commonly-received theory of gravity, accompanied with a declaration of Kepler's own opinions on the same subject. Some of the most remarkable passages in it have been already quoted in the life of Galileo; but, nevertheless, they are too important to Kepler's reputation to be omitted here, containing as they do a distinct and positive enunciation of the law of universal gravitation. It does not appear, however, that Kepler estimated rightly the importance of the theory here traced out by him, since on every other occasion he advocated principles with which it is scarcely reconcileable. The discussion is introduced in the following terms:—

"The motion of heavy bodies hinders many from believing that the earth is moved by an animal motion, or rather a magnetic one. Let such consider the following propositions. A mathematical point, whether the centre of the universe or not, has no power, either effectively or objectively, to move heavy bodies to approach it. Let physicians prove if they can, that such power can be possessed by a point, which neither is a body, nor is conceived unless by relation alone. It is impossible that the form[*] of a stone should, by moving its own body, seek a mathematical point, or in other words, the centre of the universe, without regard of the body in which that point exists. Let physicians prove if they can, that natural things have any sympathy with that which is nothing. Neither do heavy bodies tend to the centre of the universe by reason that they are avoiding the extremities of the round universe; for their distance from the centre is insensible, in proportion to their distance from the extremities of the universe. And what reason could there be for this hatred? How strong, how wise must those heavy bodies be, to be able to escape so carefully from an enemy lying on all sides of

---
[*] It is not very easy to carry the understanding aright among these Aristotelian ideas. Many at the present day might think they understood better what is meant, if for " form" had been written " nature."

them: what activity in the extremities of the world to press their enemy so closely! Neither are heavy bodies driven into the centre by the whirling of the first moveable, as happens in revolving water. For if we assume such a motion, either it would not be continued down to us, or otherwise we should feel it, and be carried away with it, and the earth also with us; nay, rather, we should be hurried away first, and the earth would follow; all which conclusions are allowed by our opponents to be absurd. It is therefore plain that the vulgar theory of gravity is erroneous.

The true theory of gravity is founded on the following axioms:—Every corporeal substance, so far forth as it is corporeal, has a natural fitness for resting in every place where it may be situated by itself beyond the sphere of influence of a body cognate with it. Gravity is a mutual affection between cognate bodies towards union or conjunction (similar in kind to the magnetic virtue), so that the earth attracts a stone much rather than the stone seeks the earth. Heavy bodies (if we begin by assuming the earth to be in the centre of the world) are not carried to the centre of the world in its quality of centre of the world, but as to the centre of a cognate round body, namely, the earth; so that wheresoever the earth may be placed, or whithersoever it may be carried by its animal faculty, heavy bodies will always be carried towards it. If the earth were not round, heavy bodies would not tend from every side in a straight line towards the centre of the earth, but to different points from different sides. If two stones were placed in any part of the world near each other, and beyond the sphere of influence of a third cognate body, these stones, like two magnetic needles, would come together in the intermediate point, each approaching the other by a space proportional to the comparative mass of the other. If the moon and earth were not retained in their orbits by their animal force or some other equivalent, the earth would mount to the moon by a fifty-fourth part of their distance, and the moon fall towards the earth through the other fifty-three parts and they would there meet; assuming however that the substance of both is of the same density. If the earth should cease to attract its waters to itself, all the waters of the sea would be raised and would flow to the body of the moon. The sphere of the attractive virtue which is in the moon extends as far as the earth, and entices up the waters; but as the moon flies rapidly across the zenith, and the waters cannot follow so quickly, a flow of the ocean is occasioned in the torrid zone towards the westward. If the attractive virtue of the moon extends as far as the earth, it follows with greater reason that the attractive virtue of the earth extends as far as the moon, and much farther; and in short, nothing which consists of earthly substance any how constituted, although thrown up to any height, can ever escape the powerful operation of this attractive virtue. Nothing which consists of corporeal matter is absolutely light, but that is comparatively lighter which is rarer, either by its own nature, or by accidental heat. And it is not to be thought that light bodies are escaping to the surface of the universe while they are carried upwards, or that they are not attracted by the earth. They are attracted, but in a less degree, and so are driven outwards by the heavy bodies; which being done, they stop, and are kept by the earth in their own place. But although the attractive virtue of the earth extends upwards, as has been said, so very far, yet if any stone should be at a distance great enough to become sensible, compared with the earth's diameter, it is true that on the motion of the earth such a stone would not follow altogether; its own force of resistance would be combined with the attractive force of the earth, and thus it would extricate itself in some degree from the motion of the earth."

Who, after perusing such passages in the works of an author, whose writings were in the hands of every student of astronomy, can believe that Newton waited for the fall of an apple to set him thinking for the first time on the theory which has immortalized his name? An apple may have fallen, and Newton may have seen it; but such speculations as those which it is asserted to have been the cause of originating in him had been long familiar to the thoughts of every one in Europe pretending to the name of natural philosopher.

As Kepler always professed to have derived his notion of a magnetic attraction among the planetary bodies from the writings of Gilbert, it may be worth while to insert here an extract from the "New Philosophy" of that author, to show in what form he presented a similar theory of the tides, which affords the most striking illustration of that attraction. This work was not published till the middle of the seventeenth century, but a knowledge of its contents may, in several instances, be traced back to the period in which it was written:—

"There are two primary causes of the motion of the seas—the moon, and the diurnal revolution. The moon does not act on the seas by its rays or its light. How then? Certainly by the common effort of the bodies, and (to explain it by something similar) by their magnetic attraction. It should be known, in the first place, that the whole quantity of water is not contained in the sea and rivers, but that the mass of earth (I mean this globe) contains moisture and spirit much deeper even than the sea. The moon draws this out by sympathy, so that they burst forth on the arrival of the moon, in consequence of the attraction of that star; and for the same reason, the quicksands which are in the sea open themselves more, and perspire their moisture and spirits during the flow of the tide, and the whirlpools in the sea disgorge copious waters; and as the star retires, they devour the same again, and attract the spirits and moisture of the terrestrial globe. Hence the moon attracts, not so much the sea as the subterranean spirits and humours; and the interposed earth has no more power of resistance than a table or any other dense body has to resist the force of a magnet. The sea rises from the greatest depths, in consequence of the ascending humours and spirits; and when it is raised up, it necessarily flows on to the shores, and from the shores it enters the rivers."*

This passage sets in the strongest light one of the most notorious errors of the older philosophy, to which Kepler himself was remarkably addicted. If Gilbert had asserted, in direct terms, that the moon attracted the water, it is certain that the notion would have been stigmatized (as it was for a long time in Newton's hands) as arbitrary, occult, and unphilosophical: the idea of these subterranean humours was likely to be treated with much more indulgence. A simple statement, that when the moon was over the water the latter had a tendency to rise towards it, was thought to convey no instruction; but the assertion that the moon draws out subterranean spirits by sympathy, carried with it a more imposing appearance of theory. The farther removed these humours were from common experience, the easier it became to discuss them in vague and general language; and those who called themselves philosophers could endure to hear attributes bestowed on these fictitious elements which revolted their imaginations when applied to things of whose reality at least some evidence existed.

It is not necessary to dwell upon the system of Tycho Brahe, which was identical, as we have said, with one rejected by Copernicus, and consisted in making the sun revolve about the earth, carrying with it all the other planets revolving about him. Tycho went so far as to deny the rotation of the earth to explain the vicissitudes of day and night, but even his favourite assistant Longomontanus differed from him in this part of his theory. The great merit of Tycho Brahe, and the service he rendered to astronomy, was entirely independent of any theory; consisting in the vast accumulation of observations made by him during a residence of fifteen years at Uraniburg, with the assistance of instruments, and with a degree of care, very far superior to anything known before his time in practical astronomy. Kepler is careful repeatedly to remind us, that without Tycho's observations he could have done nothing. The degree of reliance that might be placed on the results obtained by observers who acknowledged their inferiority to Tycho Brahe, may be gathered from an incidental remark of Kepler to Longomontanus. He had been examining Tycho's registers, and had occasionally found a difference amounting sometimes to 4' in the right ascensions of same planet, deduced from different stars on the same night. Longomontanus could not deny the fact, but declared that it was impossible to be always correct within such limits. The reader should never lose sight of this uncertainty in the observations, when endeavouring to estimate the difficulty of finding a theory that would properly represent them.

When Kepler first joined Tycho Brahe at Prague, he found him and Longomontanus very busily engaged in correcting the theory of Mars, and accordingly it was this planet to which he also first directed his attention. They had formed a catalogue of the mean oppositions of Mars during twenty years, and had discovered a position of the equant, which (as they said) represented them with tolerable

* 'De mundo nostro sublunari, Philosophia Nova. Amstelodami, 1651.

exactness. On the other hand, they were much embarrassed by the unexpected difficulties they met in applying a system which seemed on the one hand so accurate, to the determination of the latitudes, with which it could in no way be made to agree. Kepler had already suspected the cause of this imperfection, and was confirmed in the view he took of their theory, when, on a more careful examination, he found that they overrated the accuracy even of their longitudes. The errors in these, instead of amounting as they said, nearly to 2', rose sometimes above 21'. In fact they had reasoned ill on their own principles, and even if the foundations of their theory had been correctly laid, could not have arrived at true results. But Kepler had satisfied himself of the contrary, and the following diagram shews the nature of the first alteration he introduced, not perhaps so celebrated as some of his later discoveries, but at least of equal consequence to astronomy, which could never have been extricated from the confusion into which it had fallen, till this important change had been effected.

The practice of Tycho Brahe, indeed of all astronomers till the time of Kepler, had been to fix the position of the planet's orbit and equant from observations on its mean oppositions, that is to say, on the times when it was precisely six signs or half a circle distant from the mean place of the sun. In the annexed figure, let S represent the sun, C the centre of the earth's orbit, T t.

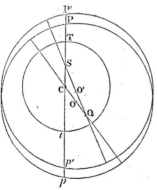

Tycho Brahe's practice amounted to this, that if Q were supposed the place of the centre of the planet's equant, the centre of P p its orbit was taken in Q C, and not in Q S, as Kepler suggested that it ought to be taken. The consequence of this erroneous practice was, that the observa-

tions were deprived of the character for which oppositions were selected, of being entirely free from the second inequalities. It followed therefore that as part of the second inequalities were made conducive towards fixing the relative position of the orbit and equant, to which they did not naturally belong, there was an additional perplexity in accounting for the remainder of them by the size and motion of the epicycle. As the line of nodes of every planet was also made to pass through C instead of S, there could not fail to be corresponding errors in the latitudes. It would only be in the rare case of an opposition of the planet in the line C S, that the time of its taking place would be the same, whether O, the centre of the orbit, was placed in C Q or S Q. Every other opposition would involve an error, so much the greater as it was observed at a greater distance from the line C S.

It was long however before Tycho Brahe could be made to acquiesce in the propriety of the proposed alteration; and, in order to remove his doubts as to the possibility that a method could be erroneous which, as he still thought, had given him such accurate longitudes, Kepler undertook the ungrateful labour of the first part of his "Commentaries." He there shewed, in the three systems of Copernicus, Tycho Brahe, and Ptolemy, and in both the concentric and excentric theories, that though a false position were given to the orbit, the longitudes of a planet might be so represented, by a proper position of the centre of the equant, as never to err in oppositions above 5' from those given by observation; though the second inequalities and the latitudes would thereby be very greatly deranged.

The change Kepler introduced, of observing apparent instead of mean oppositions, made it necessary to be very accurate in his reductions of the planet's place to the ecliptic; and in order to be able to do this, a previous knowledge of the parallax of Mars became indispensable. His next labour was therefore directed to this point; and finding that the assistants to whom Tycho Brahe had previously committed this labour had performed it in a negligent and imperfect manner, he began afresh with Tycho's original observations. Having satisfied himself as to the probable limits of his errors in the parallax on which he finally fixed, he proceeded to determine the inclination of the orbit and the position of the line of nodes. In all these operations his talent for astronomical inquiries appeared pre-eminent in a variety of new methods by which he combined and availed himself of the observations; but it must be sufficient merely to mention this fact, without entering into any detail. One important result may be mentioned, at which he arrived in the course of them, the constancy of the inclination of the planet's orbit, which naturally strengthened him in his new theory.

Having gone through these preliminary inquiries, he came at last to fix the proportions of the orbit; and, in doing so, he determined, in the first instance, not to assume, as Ptolemy appeared to have done arbitrarily, the bisection of the excentricity, but to investigate its proportion along with the other elements of the orbit, which resolution involved him in much more laborious calculations. After he had gone over all the steps of his theory no less than seventy times—an appalling labour, especially if we remember that logarithms were not then invented—his final result was, that in 1587, on the 6th of March, at 7^h $23'$, the longitude of the aphelion of Mars was 4^s $28°$ $48'$ $55''$; that the planet's mean longitude was 6^s $0°$ $51'$ $35''$; that if the semidiameter of the orbit was taken at 100000, the excentricity was 11332; and the excentricity of the equant 18564. He fixed the radius of the greater epicycle at 14988, and that of the smaller at 3628.

When he came to compare the longitudes as given by this, which he afterwards called the *vicarious* theory, with the observations at opposition, the result seemed to promise him the most brilliant success. His greatest error did not exceed $2'$; but, notwithstanding these flattering anticipations, he soon found by a comparison of longitudes out of opposition and of latitudes, that it was yet far from being so complete as he had imagined, and to his infinite vexation he soon found that the labour of four years, which he had expended on this theory, must be considered almost entirely fruitless. Even his favourite principle of dividing the excentricity in a different ratio from Ptolemy, was found to lead him into greater error than if he had retained the old bisection. By restoring that, he made his latitudes more accurate, but produced a corresponding change for the worse in his longitudes; and although the errors of $8'$, to which they now amounted, would probably have been disregarded by former theorists, Kepler could not remain satisfied till they were accounted for. Accordingly he found himself forced to the conclusion that one of the two principles on which this theory rested must be erroneous; either the orbit of the planet is not a perfect circle, or there is no fixed point within it round which it moves with an uniform angular motion. He had once before admitted the possibility of the former of these facts, conceiving it possible that the motion of the planets is not at all curvilinear, but that they move in polygons round the sun, a notion to which he probably inclined in consequence of his favourite harmonics and geometrical figures.

In consequence of the failure of a theory conducted with such care in all its practical details, Kepler determined that his next trial should be of an entirely different complexion. Instead of first satisfying the first inequalities of the planet, and then endeavouring to account for the second inequalities, he resolved to reverse the process, or, in other words, to ascertain as accurately as possible what part of the planet's apparent motion should be referred solely to the optical illusion produced by the motion of the earth, before proceeding to any inquiry of the real inequality of the planet's proper motion. It had been hitherto taken for granted, that the earth moved equably round the centre of its orbit; but Kepler, on resuming the consideration of it, recurred to an opinion he had entertained very early in his astronomical career (rather from his conviction of the existence of general laws, than that he had then felt the want of such a supposition), that it required an equant distinct from its orbit no less than the other planets. He now saw, that if this were admitted, the changes it would everywhere introduce in the optical part of the planet's irregularities might perhaps relieve him from the perplexity in which the vicarious theory had involved him. Accordingly he applied himself with renewed assiduity to the examination of this important question, and the result of his calculations (founded principally on observations of Mars' parallax) soon satisfied him not only that the earth's orbit does require such an equant, but that its centre is placed according to the general law of the bisection of the excentricity which he had previously found indispensable in the other planets. This was an innovation of the first magnitude, and accordingly Kepler did not venture to proceed farther in his theory, till by evidence of the most varied and satisfactory nature, he had established it beyond the possibility of cavil.

It may be here remarked, that this principle of the bisection of the eccentricity, so familiar to the Ptolemaic astronomers, is identical with the theory afterwards known by the name of the simple elliptic hypothesis, advocated by Seth Ward and others. That hypothesis consisted in supposing the sun to be placed in one focus of the elliptic orbit of the planet, whose angular motion was uniform round the other focus. In Ptolemaic phraseology, that other focus was the centre of the equant, and it is well known that the centre of the ellipse lies in the middle point between the two foci.

It was at this period also, that Kepler first ventured upon the new method of representing inequalities which terminated in one of his most celebrated discoveries. We have already seen, in the account of the "Mysterium Cosmographicum," that he was speculating, even at that time, on the effects of a whirling force exerted by the sun on the planets with diminished energy at increased distances, and on the proportion observed between the distances of the planets from the sun, and their periods of revolution. He seems even then to have believed in the possibility of discovering a relation between the times and distances in different planets. Another analogous consequence of his theory of the radiation of the whirling force would be, that if the same planet should recede to a greater distance from the central body, it would be acted on by a diminished energy of revolution, and consequently, a relation might be found between the velocity at any point of its orbit, and its distance at that point from the sun. Hence he expected to derive a more direct and natural method of calculating the inequalities, than from the imaginary equant. But these ingenious ideas had been checked in the outset by the erroneous belief which Kepler, in common with other astronomers, then entertained of the coincidence of the earth's equant with its orbit; in other words, by the belief that the earth's linear motion was uniform, though it was known not to remain constantly at the same distance from the sun. As soon as this prejudice was removed, his former ideas recurred to him with increased force, and he set himself diligently to consider what relation could be found between the velocity and distance of a planet from the sun. The method he adopted in the beginning of this inquiry was to assume as approximately correct Ptolemy's doctrine of the bisection of the excentricity, and to investigate some simple relation nearly representing the same effect.

In the annexed figure, S is the place of the sun, C the centre of the planet's

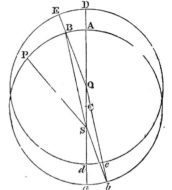

orbit $A B a b$, Q the centre of the equant represented by the equal circle $D E d e$. $A B$, $a b$, two equal small arcs described by the planet at the apsides of its orbit: then, according to Ptolemy's principles, the arc $D E$ of the equant would be proportional to the time of passing along $A B$, on the same scale on which $d e$ would represent the time of passing through the equal arc $a b$.

$Q D : Q A : : D E : A B$, nearly; and because $Q S$ is bisected in C, $Q A$, $C A$ or $Q D$, and $S A$, are in arithmetical proportion: and, therefore, since an arithmetical mean, when the difference is small, does not differ much from a geometrical mean, $Q D : Q A : : S A : Q D$, nearly. Therefore, $D E : A B : : S A : Q D$, nearly, and in the same manner $d e : a b : : S a : Q d$ nearly; and therefore $D E : d e : : S A : S a$ nearly. Therefore at the apsides, the times of passing over equal spaces, on Ptolemy's theory, are nearly as the distances from the sun, and Kepler, with his usual hastiness, immediately concluded that this was the accurate and general law, and that the errors of the old theory arose solely from having departed from it.

It followed immediately from this assumption, that after leaving the point A, the time in which the planet would arrive at any point P of its orbit would be proportional to, and might be represented by, the sums of all the lines that could be drawn from S to the arc A P, on the same scale that the whole period of revolution would be denoted by the sum of all the lines drawn to every point of the orbit. Kepler's first attempt to verify this supposition approximately, was made by dividing the whole circumference of the orbit into 360 equal parts, and calculating the distances at every one of the points of division. Then supposing the planet to move uniformly, and to remain at the same distance from the sun during the time of passing each one of these divisions, (a supposition which manifestly would not differ much from the former one, and would coincide with it more nearly, the greater was the number of divisions taken) he proceeded to add together these calculated distances, and hoped to find that the time of arriving at any one of the divisions bore the same ratio to the whole period, as the sum of the corresponding set of distances did to the sum of the whole 360.

This theory was erroneous; but by almost miraculous good fortune, he was led by it in the following manner to the true measure. The discovery was a consequence of the tediousness of his first method, which required, in order to know the time of arriving at any point, that the circle should be subdivided, until one of the points of division fell exactly upon the given place. Kepler therefore endeavoured to discover some shorter method of representing these sums of the distances. The idea then occurred to him of employing for that purpose the area inclosed between the two distances, S A, S P, and the arc A P, in imitation of the manner in which he remembered that Archimedes had found the area of the circle, by dividing it into an infinite number of small triangles by lines drawn from the centre. He hoped therefore to find, that the time of passing from A to P bore nearly the same ratio to the whole period of revolution that the area A S P bore to the whole circle.

This last proportion is in fact accurately observed in the revolution of one body round another, in consequence of an attractive force in the central body. Newton afterwards proved this, grounding his demonstration upon laws of motion altogether irreconcileable with Kepler's opinions; and it is impossible not to admire Kepler's singular good fortune in arriving at this correct result in spite, or rather through the means, of his erroneous principles. It is true that the labour which he bestowed unsparingly upon every one of his successive guesses, joined with his admirable candour, generally preserved him from long retaining a theory altogether at variance with observations; and if any relation subsisted between the times and distances which could any way be expressed by any of the geometrical quantities under consideration, he could scarcely have failed—it might be twenty years earlier or twenty years later,—to light upon it at last, having once put his indefatigable fancy upon this scent. But in order to prevent an over-estimate of his merit in detecting this beautiful law of nature, let us for a moment reflect what might have been his fate had he endeavoured in the same manner, and with the same perseverance, to discover a relation, where, in reality, none existed. Let us take for example the inclinations or the excentricities of the planetary orbits, among which no relation has yet been discovered; and if any exists, it is probably of too complicated a nature to be hit at a venture. If Kepler had exerted his ingenuity in this direction, he might have wasted his life in fruitless labour, and whatever reputation he might have left behind him as an industrious calculator, it would have been very far inferior to that which has procured for him the proud title of the " Legislator of the Heavens."

However this may be, the immediate consequence of thus lighting upon the real law observed by the earth in its passage round the sun was, that he found himself in possession of a much more accurate method of representing its inequalities than had been reached by any of his predecessors; and with renewed hopes he again attacked the planet Mars, whose path he was now able to consider undistorted by the illusions arising out of the motion of the earth. Had the path of Mars been accurately circular, or even as nearly approaching a circle as that of the earth, the method he chose of determining its position and size by means of three distances carefully calculated from his observed parallaxes, would have given a satisfactory result; but finding, as he soon did, that almost every set of three distances led him to a different result, he began to suspect another error in the long-received opinion, that the orbits of the planets must consist of a combination of circles; he therefore determined, in the first instance, to fix the distances of the planet at the apsides without any reference to the form of the intermediate orbit. Half the difference between these would, of course, be the excentricity of the orbit; and as this quantity came out very nearly the same as had been determined on the vicarious theory, it seemed clear that the error of that theory, whatever it might be, did not lie in these elements.

Kepler also found that in the case of this planet likewise, the times of describing equal arcs at the apsides were proportional to its distances from the sun, and he naturally expected that the method of areas would measure the planet's motion with as much accuracy as he had found in the case of the earth. This hope was disappointed: when he calculated the motion of the planet by this method, he obtained places too much advanced when near the apsides, and too little advanced at the mean distances. He did not, on that account, immediately reject the opinion of circular orbits, but was rather inclined to suspect the principle of measurement, at which he felt that he had arrived in rather a precarious manner. He was fully sensible that his areas did not accurately represent the sums of any distances except those measured from the centre of the circle; and for some time he abandoned the hope of being able to use this substitution, which he always considered merely as an approximate representation of the true measure, the sum of the distances. But on examination he found that the errors of this substitution were nearly insensible, and those it did in fact produce, were in the contrary direction of the errors he was at this time combating. As soon as he had satisfied himself of this, he ventured once more on the supposition, which by this time had, in his eyes, almost acquired the force of demonstration, that the orbits of the planets are not circular, but of an oval form, retiring within the circle at the mean distances, and coinciding with it at the apsides.

This notion was not altogether new; it had been suggested in the case of Mercury, by Purbach, in his "Theories of the Planets." In the edition of this work published by Reinhold, the pupil of Copernicus, we read the following passage. "Sixthly, it appears from what has been said, that the centre of Mercury's epicycle, by reason of the motions above-mentioned, does not, as is the case with the other planets, describe the circumference of a circular deferent, but rather the periphery of a figure resembling a plane oval." To this is added the following note by Reinhold. " The centre of the Moon's epicycle describes a path of a lenticular shape; Mercury's on the contrary is egg-shaped, the big end lying towards his apogee, and the little end towards his perigee*." The excentricity of Mercury's orbit is, in fact, much greater than that of any of the other planets, and the merit of making this first step cannot reasonably be withheld from Purbach and his commentator, although they did not pursue the inquiry so far as Kepler found himself in a condition to do.

Before proceeding to the consideration of the particular oval which Kepler fixed upon in the first instance, it will be necessary, in order to render intelligible the source of many of his doubts and difficulties, to make known something more of his theory of the moving force by which he supposed the planets to be carried round in their orbits. In conformity with the plan hitherto pursued, this shall be done as much as possible in his own words.

" It is one of the commonest axioms in natural philosophy, that if two things always happen together and in the same manner, and admit the same measure, either the one is the cause of the other, or both are the effect of a common cause. In the present case, the increase or languor of motion invariably corresponds with an approach to or departure from the centre of the universe. Therefore, either the languor is the cause of the departure of the star, or the departure of the languor, or both have a common cause. But no one can be of opinion that there is a concurrence of any third thing to be a common cause of these two effects, and in the following chapters it will be made clear that there is no occasion to imagine any such third thing, since the two are of themselves sufficient. Now, it is not agreeable to the nature of things that activity or languor in linear motion should be the cause of distance from the centre. For, distance from the centre is conceived anteriorly to linear motion. In fact linear motion cannot exist without dis-

---

* Theoricæ novæ planetarum. G. Purbachii, Parisiis, 1553.

tance from the centre, since it requires space for its accomplishment, but distance from the centre can be conceived without motion. Therefore distance is the cause of the activity of motion, and a greater or less distance of a greater or less delay. And since distance is of the kind of relative quantities, whose essence consists in boundaries, (for there is no efficacy in relation *per se* without regard to bounds,) it follows that the cause of the varying activity of motion rests in one of the boundaries. But the body of the planet neither becomes heavier by receding, nor lighter by approaching. Besides, it would perhaps be absurd on the very mention of it, that an animal force residing in the moveable body of the planet for the purpose of moving it, should exert and relax itself so often without weariness or decay. It remains, therefore, that the cause of this activity and languor resides at the other boundary, that is, in the very centre of the world, from which the distances are computed.— Let us continue our investigation of this moving virtue which resides in the sun, and we shall presently recognize its very close analogy to light. And although this moving virtue cannot be identical with the light of the sun, let others look to it whether the light is employed as a sort of instrument, or vehicle, to convey the moving virtue. There are these seeming contradictions:—first, light is obstructed by opaque bodies, for which reason if the moving virtue travelled on the light, darkness would be followed by a stoppage of the moveable bodies. Again, light flows out in right lines spherically, the moving virtue in right lines also, but cylindrically: that is, it turns in one direction only, from west to east; not in the opposite direction, not towards the poles, &c. But perhaps we shall be able presently to reply to these objections. In conclusion, since there is as much virtue in a large and remote circle as in a narrow and close one, nothing of the virtue perishes in the passage from its source, nothing is scattered between the source and the moveable. Therefore the efflux, like that of light, is not material, and is unlike that of odours, which are accompanied by a loss of substance, unlike heat from a raging furnace, unlike every other emanation by which mediums are filled. It remains, therefore, that as light which illuminates all earthly things, is the immaterial species of that fire which is in the body of the sun, so this virtue, embracing and moving all the planetary bodies, is the immaterial species of that virtue which resides in the sun itself, of incalculable energy, and so the primary act of all mundane motion.—I should like to know who ever said that there was anything material in light!—Guided by our notion of the efflux of this species (or archetype), let us contemplate the more intimate nature of the source itself. For it seems as if something divine were latent in the body of the sun, and comparable to our own soul, whence that species emanates which drives round the planets; just as from the mind of a slinger the species of motion sticks to the stones, and carries them forward, even after he who cast them has drawn back his hand. But to those who wish to proceed soberly, reflections differing a little from these will be offered."

Our readers will, perhaps, be satisfied with the assurance, that these sober considerations will not enable them to form a much more accurate notion of Kepler's meaning than the passages already cited. We shall therefore proceed to the various opinions he entertained on the motion of the planets.

He considered it as established by his theory, that the centre E of the planet's epicycle (see fig. p. 33.) moved round the circumference of the deferent Dd, according to the law of the planet's distances; the point remaining to be settled was the motion of the planet in the epicycle. If it were made to move according to the same law, so that when the centre of the epicycle reached E, the planet should be at F, taking the angle BEF equal to BSA, it has been shewn (p. 19) that the path of F would still be a circle, excentric from Dd by DA the radius of the epicycle.

But Kepler fancied that he saw many sound reasons why this could not be the true law of motion in the epicycle, on which reasons he relied much more firmly than on the indisputable fact, which he mentions as a collateral proof, that it was contradicted by the observations. Some of these reasons are subjoined: " In the beginning of the work it has been declared to be most absurd, that a planet (even though we suppose it endowed with mind) should form any notion of a centre, and a distance from it, if there be no body in that centre to serve for a distinguishing mark. And although you should say, that the planet has respect to the sun, and knows beforehand, and remembers the order in which the distances from the sun are comprised, so as to make a perfect excentric; in the first place, this is rather far-fetched, and requires, in any mind, means for connecting the effect of an accurately circular path with the sign of an increasing and diminishing diameter of the sun. But there are no such means, except the position of the centre of the excentric at a given distance from the sun; and I have already said, that this is beyond the power of a mere mind. I do not deny that a centre may be imagined, and a circle round it; but this I do say, if the circle exists only in imagination, with no external sign or division, that it is not possible that the path of a moveable body should be really ordered round it in an exact circle. Besides, if the planet chooses from memory its just distances from the sun, so as exactly to form a circle, it must also take from the same source, as if out of the Prussian or Alphonsine tables, equal excentric arcs, to be described in unequal times, and to be described by a force extraneous from the sun; and thus would have, from its memory, a foreknowledge of what effects a virtue, senseless and extraneous from the sun, was about to produce: all these consequences are absurd."

"It is therefore more agreeable to reason that the planet takes no thought, either of the excentric or epicycle; but that the work which it accomplishes, or joins in effecting, is a libratory path in the diameter B b of the epicycle, in the direction towards the sun. The law is now to be discovered, according to which the planet arrives at the proper distances in any time. And indeed in this inquiry, it is easier to say what the law is not than what it is."—Here, according to his custom, Kepler enumerates several laws of motion by which the planet might choose to regulate its energies, each of which is successively condemned. Only one of them is here mentioned, as a specimen of the rest. "What then if we were to say this? Although the motions of the planet are not epicyclical, perhaps the libration is so arranged that the distances from the sun are equal to what they would have been in a real epicyclical motion.—This leads to more incredible consequences than the former suppositions, and yet in the dearth of better opinions, let us for the present content ourselves with this. The greater number of absurd conclusions it will be found to involve, the more ready will a physician be, when we come to the fifty-second chapter, to admit what the observations testify, that the path of the planet is not circular."

The first oval path on which Kepler was induced to fix, by these and many other similar considerations, was in the first instance very different from the true elliptical form. Most authors would have thought it unnecessary to detain their readers with a theory which they had once entertained and rejected; but Kepler's work was written on a different plan. He thus introduces an explanation of his first oval. "As soon as I was thus taught by Brahe's very accurate observations that the orbit of a planet is not circular, but more compressed at the sides, on the instant I thought that I understood the natural cause of this deflection. But the old proverb was verified in my case;—the more haste the less speed.—For having violently laboured in the 39th chapter, in consequence of my inability to find a sufficiently probable cause why the orbit of the planet should be a perfect circle, (some absurdities always remaining with respect to that virtue which resides in the body of the planet,) and having now discovered from the observations, that the orbit is not a perfect circle, I felt furiously inclined to believe that if the theory which had been recognized as absurd, when employed in the 39th chapter for the purpose of fabricating a circle, were modulated into a more probable form, it would produce an accurate orbit agreeing with the observations. If I had entered on this course a little more warily, I might have detected the truth immediately. But, being blinded by my eagerness, and not sufficiently regardful of every part of the 39th chapter, and clinging to my first opinion, which offered itself to me with a wonderful show of probability, on account of the equable motion in the epicycle, I got entangled in new perplexities, with which we shall now have to struggle in this 45th chapter and the following ones as far as the 50th chapter."

In this theory, Kepler supposed that whilst the centre of the epicycle was moving round a circular deferent according to the law of the planets' distances (or areas) the planet itself moved equably in the epicycle, with the mean angular velocity of its centre in the deferent. In consequence of this supposition, since

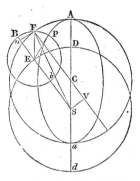

at D, when the planet is at A the aphelion, the motion in the deferent is less than the mean motion, the planet will have advanced through an angle B E P greater than B E F or B S A, through which the centre of the epicycle has moved; and consequently, the path will lie everywhere within the circle A a, except at the apsides. Here was a new train of laborious calculations to undergo for the purpose of drawing the curve A P a according to this law, and of measuring the area of any part of it. After a variety of fruitless attempts, for this curve is one of singular complexity, he was reduced, as a last resource, to suppose it insensibly different from an ellipse on the same principal axes, as an approximate means of estimating its area. Not content even with the results so obtained, and not being able to see very clearly what might be the effect of his alteration in substituting the ellipse for the oval, and in other simplifications introduced by him, he had courage enough to obtain the sums of the 360 distances by direct calculation, as he had done in the old circular theory.

In the preface to his book he had spoken of his labours under the allegory of a war carried on by him against the planet; and when exulting in the early prospects of success this calculation seemed to offer, he did not omit once more to warn his readers, in his peculiar strain, that this exultation was premature.

"'Allow me, gentle reader, to enjoy so splendid a triumph for one little day (I mean through the five next chapters), meantime be all rumours suppressed of new rebellion, that our preparations may not perish, yielding us no delight. Hereafter if anything shall come to pass, we will go through it in its own time and season; now let us be merry, as then we will be bold and vigorous." At the time foretold, that is to say, at the end of the five merry chapters, the bad news could no longer be kept a secret. It is announced in the following bulletin:—
" While thus triumphing over Mars, and preparing for him, as for one altogether vanquished, tabular prisons, and equated eccentric fetters, it is buzzed here and there that the victory is vain, and that the war is raging anew as violently as before. For the enemy, left at home a despised captive, has burst all the chains of the equations, and broken forth of the prisons of the tables. For no method of geometrically administering the theory of the 45th chapter was able to come near the accuracy of approximation of the vicarious theory of the 16th chapter, which gave me true equations derived from false principles. Skirmishers, disposed all round the circuit of the excentric, (I mean the true distances,) routed my forces of physical causes levied out of the 45th chapter, and shaking off the yoke, regained their liberty. And now there was little to prevent the fugitive enemy from effecting a junction with his rebellious supporters, and reducing me to despair, had I not suddenly sent into the field a reserve of new physical reasonings on the rout and dispersion of the veterans, and diligently followed, without allowing him the slightest respite, in the direction in which he had broken out."

In plainer terms, Kepler found, after this labour was completed, that the errors in longitude he was still subject to were precisely of an opposite nature to those he had found with the circle; instead of being too quick at the apsides, the planet was now too slow there, and too much accelerated in the mean distances; and the distances obtained from direct observation were everywhere greater, except at the apsides, than those furnished by this oval theory. It was in the course of these tedious investigations that he established, still more satisfactorily than he had before done, that the inclinations of the planets' orbits are invariable, and that the lines of their nodes pass through the centre of the Sun, and not, as before his time had been supposed, through the centre of the ecliptic.

When Kepler found with certainty that this oval from which he expected so much would not satisfy the observations, his vexation was extreme, not merely from the mortification of finding a theory confuted on which he had spent such excessive labour, for he was accustomed to disappointments of that kind, but principally from many anxious and fruitless speculations as to the real physical causes why the planet did not move in the supposed epicycle, that being the point of view, as has been already shewn, from which he always preferred to begin his inquiries. One part of the reasoning by which he reconciled himself to the failure exhibits much too curious a view of the state of his mind to be passed over in silence. The argument is founded on the difficulty which he met with, as abovementioned, in calculating the proportions of the oval path he had imagined. "In order that you may see the cause of the impracticability of this method which we have just gone through, consider on what foundations it rests. The planet is supposed to move equably in the epicycle, and to be carried by the Sun unequably in the proportion of the distances. But by this method it is impossible to be known how much of the oval path corresponds to any given time, although the distance at that part is known, unless we first know the length of the whole oval. But the length of the oval cannot be known, except from the law of the entry of the planet within the sides of the circle. But neither can the law of this entry be known before we know how much of the oval path corresponds to any given time. Here you see that there is a *petitio principii*; and in my operations I was assuming that of which I was in search, namely, the length of the oval. This is at least not the fault of my understanding, but it is also most alien to the primary Ordainer of the planetary courses: I have never yet found so ungeometrical a contrivance in his other works. Therefore we must either hit upon some other method of reducing the theory of the 45th chapter to calculation; or if that cannot be done, the theory itself, suspected on account of this *petitio principii*, will totter." Whilst his mind was thus occupied, one of those extraordinary accidents which it has been said never occur but to those capable of deriving advantage from them (but which, in fact, are never noticed when they occur to any one else), fortunately put him once more upon the right path. Half the extreme breadth between the oval and the circle nearly represented the errors of his distances at the mean point, and he found that this half was 429 parts of a radius, consisting of 100000 parts;

and happening to advert to the greatest optical inequality of Mars, which amounts to about 5° 18′, it struck him that 429 was precisely the excess of the secant of 5° 18′ above the radius taken at 100000. This was a ray of light, and, to use his own words, it roused him as out of sleep. In short, this single observation was enough to produce conviction in his singularly constituted mind, that instead of the distances S F, he should everywhere substitute F V, determined by drawing S V perpendicular on the line F C, since the excess of S F above F V is manifestly that of the secant above the radius in the optical equation S F C at that point. It is still more extraordinary that a substitution made for such a reason should have the luck, as is again the case, to be the right one. This substitution in fact amounted to supposing that the planet, instead of being at the distance S P or S F, was at S n; or, in other words, that instead of revolving in the circumference, it librated in the diameter of the epicycle, which was to him an additional recommendation. Upon this new supposition a fresh set of distances was rapidly calculated, and to Kepler's inexpressible joy, they were found to agree with the observations within the limits of the errors to which the latter were necessarily subject: Notwithstanding this success, he had to undergo, before arriving at the successful termination of his labours, one more disappointment. Although the distance corresponding to a time from the aphelion represented approximately by the area A S F, was thus found to be accurately represented by the line S n, there was still an error with regard to the direction in which that distance was to be measured. Kepler's first idea was to set it off in the direction S F, but this he found to lead to inaccurate longitudes;

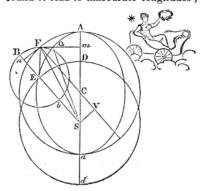

and it was not until after much perplexity, driving him, as he tells us, "almost to insanity," that he satisfied himself that the distance S Q equal to F V ought to be taken terminating in F m, the line from F perpendicular to A a, the line of apsides, and that the curve so traced out by Q would be an accurate ellipse.

He then found to his equal gratification and amazement, a small part of which he endeavoured to express by a triumphant figure on the side of his diagram, that the error he had committed in taking the area A S F to represent the sums of the distances S F, was exactly counterbalanced; for this area does accurately represent the sums of the distances F V or S Q. This compensation, which seemed to Kepler the greatest confirmation of his theory, is altogether accidental and immaterial, resulting from the relation between the ellipse and circle. If the laws of planetary attraction had chanced to have been any other than those which cause them to describe ellipses, this last singular confirmation of an erroneous theory could not have taken place, and Kepler would have been forced either to abandon the theory of the areas, which even then would have continued to measure and define their motions, or to renounce the physical opinions from which he professed to have deduced it as an approximative truth.

These are two of the three celebrated theorems called Kepler's laws: the first is, that the planets move in ellipses round the sun, placed in the focus; the second, that the time of describing any arc is proportional in the same orbit to the area included between the arc and the two bounding distances from the sun. The third will be mentioned on another occasion, as it was not discovered till twelve years later. On the establishment of these two theorems, it became important to discover a method of measuring such elliptic areas, but this is a problem which cannot be accurately solved. Kepler, in offering it to the attention of geometricians, stated his belief that its solution was unattainable by direct processes, on account of the incommensurability of the arc and sine, on which the measurement of the two parts A Q m, S Q m depends. "This," says he in conclusion, "this is my belief, and whoever shall shew my mistake, and point out the true solution,

Is erit mihi magnus Apollonius."

CHAPTER VI.

Kepler appointed Professor at Linz—His second marriage—Publishes his new Method of Gauging—Refuses a Professorship at Bologna.

WHEN presenting this celebrated book to the emperor, Kepler gave notice that he contemplated a farther attack upon Mars's relations, father Jupiter, brother Mercury, and the rest; and promised that he would be successful, provided the emperor would not forget the sinews of war, and order him to be furnished anew with means for recruiting his army. The death of his unhappy patron, the Emperor Rodolph, which happened in 1612, barely in time to save him from the last disgrace of deposition from the Imperial throne, seemed to put additional difficulties in the way of Kepler's receiving the arrears so unjustly denied to him; but on the accession of Rodolph's brother, Matthias, he was again named to his post of Imperial Mathematician, and had also a permanent professorship assigned to him in the University of Linz. He quitted Prague without much regret, where he had struggled against poverty during eleven years. Whatever disinclination he might feel to depart, arose from his unwillingness to loosen still more the hold he yet retained upon the wreck of Tycho Brahe's instruments and observations. Tengnagel, son-in-law of Tycho, had abandoned astronomy for a political career, and the other members of his family, who were principally females, suffered the costly instruments to lie neglected and forgotten, although they had [1]obstructed with the utmost jealousy Kepler's attempts to continue their utility. The only two instruments Kepler possessed of his own property, were [1]" An iron sextant of $2\frac{1}{2}$ feet diameter, and a brass azimuthal quadrant, of $3\frac{1}{2}$ feet diameter, both divided into minutes of a degree." These were the gift of his friend and patron, Hoffman, the President of Styria, and with these he made all the observations which he added to those of Tycho Brahe. His constitution was not favourable to these studies, his health being always delicate, and suffering much from exposure to the night air; his eyes also were very weak, as he mentions himself in several places. In the summary of his character which he drew up when proposing to become Tycho Brahe's assistant, he describes himself as follows:—"For observations my sight is dull; for mechanical operations my hand is awkward; in politics and domestic matters my nature is troublesome and choleric; my constitution will not allow me, even when in good health, to remain a long time sedentary (particularly for an extraordinary time after dinner); I must rise often and walk about, and in different seasons am forced to make corresponding changes in my diet."

The year preceding his departure to Linz was denounced by him as pregnant with misfortune and misery. "In the first place I could get no money from the court, and my wife, who had for a long time been suffering under low spirits and despondency, was taken violently ill towards the end of 1610, with the Hungarian fever, epilepsy, and phrenitis. She was scarcely convalescent when all my three children were at once attacked with small-pox. Leopold with his army occupied the town beyond the river, just as I lost the dearest of my sons, him whose nativity you will find in my book on the new star. The town on this side of the river where I lived was harassed by the Bohemian troops, whose new levies were insubordinate and insolent: to complete the whole, the Austrian army brought the plague with them into the city. I went into Austria, and endeavoured to procure the situation which I now hold. Returning in June, I found my wife in a decline from her grief at the death of her son, and on the eve of an infectious fever; and I lost her also, within eleven days after my return. Then came fresh annoyance, of course, and her fortune was to be divided with my step-sisters. The Emperor Rodolph would not agree to my departure; vain hopes were given me of being paid from Saxony; my time and money were wasted together, till on the death of the emperor, in 1612, I was named again by his successor, and suffered to depart to Linz. These, methinks, were reasons enough why I should have overlooked not only your letters, but even astronomy itself."

Kepler's first marriage had not been a happy one; but the necessity in which he felt himself of providing some one to take charge of his two surviving children, of whom the eldest, Susanna, was born in 1602, and Louis in 1607, determined him on entering a second time into the married state. The account he has left us of the various negotiations which preceded his final choice, does not, in any point, belie the oddity of his character. His friends seem to have received a general commission to look out for a suitable match, and in a long and most amusing letter to the Baron Strahlendorf, we are made acquainted with the pretensions and qualifications of no less than eleven ladies among whom his inclinations wavered.

The first on the list was a widow, an intimate friend of his first wife's, and who, on many accounts, appeared a most eligible match. "At first she seemed favourably inclined to the proposal; it is certain that she took time to consider it, but at last she very quietly excused herself." It must have been from a recollection of this lady's good qualities that Kepler was induced to make his offer; for we learn rather unexpectedly, after being informed of her decision, that when he soon afterwards paid his respects to her, it was for the first time that he had seen her during the last six years; and he found, to his great relief, that "there was no single pleasing point about her." The truth seems to be that he was nettled by her answer, and he is at greater pains than appear necessary, considering this last discovery, to determine why she would not accept his offered hand. Among other reasons he suggested her children, among whom were two marriageable daughters; and it is diverting afterwards to find them also in the catalogue which Kepler appeared to be making of all his female acquaintance. He seems to have been much perplexed in attempting to reconcile his astrological theory with the fact of his having taken so much trouble about a negotiation not destined to succeed. " Have the stars exercised any influence here? For just about this time the direction of the Mid-Heaven is in hot opposition to Mars, and the passage of Saturn, through the ascending point of the zodiac, in the scheme of my nativity, will happen again next November and December. But if these are the causes, how do they act? Is that explanation the true one which I have elsewhere given? For I can never think of handing over to the stars the office of deities to produce effects. Let us therefore suppose it accounted for by the stars, that at this season I am violent in my temper and affections, in rashness of belief, in a shew of pitiful tenderheartedness; in catching at reputation by new and paradoxical notions, and the singularity of my actions; in busily inquiring into, and weighing and discussing, various reasons; in the uneasiness of my mind with respect to my choice. I thank God that that did not happen which might have happened; that this marriage did not take place: now for the others." Of these others, one was too old, another in bad health, another too proud of her birth and quarterings; a fourth had learned nothing but shewy accomplishments, "not at all suitable to the sort of life she would have to lead with me." Another grew impatient, and married a more decided admirer, whilst he was hesitating. "The mischief (says he) in all these attachments was, that whilst I was delaying, comparing, and balancing conflicting reasons, every day saw me inflamed with a new passion." By the time he reached the eighth, he found his match in this respect. "Fortune at length has avenged herself on my doubtful inclinations. At first she was quite complying, and her friends also: presently, whether she did or did not consent, not only I, but she herself did not know. After the lapse of a few days, came a renewed promise, which however had to be confirmed a third time; and four days after that, she again repented her confirmation, and begged to be excused from it. Upon this I gave her up, and this time all my counsellors were of one opinion." This was the longest courtship in the list, having lasted three whole months; and quite disheartened by its bad success, Kepler's next attempt was of a more timid complexion. His advances to No, 9, were made by confiding to her the whole story of his recent disappointment, prudently determining to be guided in his behaviour, by observing whether the treatment he had experienced met with a proper degree of sympathy. Apparently the experiment did not succeed; and almost reduced to despair, Kepler betook himself to the advice of a friend, who had for some time past complained that she was not consulted in this difficult negotiation. When she produced No. 10, and the first visit was paid, the report upon her was as follows:—" She has, undoubtedly, a good fortune, is of good family, and of economical habits: but her physiognomy is most horribly ugly; she would be stared at in the streets, not to mention the striking disproportion in our figures. I am lank, lean, and spare; she is short and thick: in a family notorious for fatness she is considered superfluously fat." The only objection to No. 11 seems to have been her excessive youth; and when this treaty was broken of on that account, Kepler turned his back upon all his advisers, and chose for himself one who had figured as No. 5 in the list, to whom he professes to have felt attached throughout, but from whom the represeutations of his friends had hitherto detained him, probably on account of her humble station.

The following is Kepler's summary of her character. "Her name is Susanna, the daughter of John Reuthinger and Barbara, citizens of the town of Eferdingen; the father was by trade a cabinet-maker, but both her parents are dead. She has received an education well worth the largest dowry, by favour of the Lady of Stahrenberg, the strictness of whose household is famous throughout the province. Her person and manners are suitable to mine; no pride, no extravagance; she can bear to work; she has a tolerable knowledge how to manage a family; middle-aged, and of a disposition and capability to acquire what she still wants. Her I shall marry by favour of the noble baron of Stahrenberg at twelve o'clock on the 30th of next October, with all Eferdingen assembled to meet us, and we shall eat the marriage-dinner at Maurice's at the Golden Lion."

Hantsch has made an absurd mistake with regard to this marriage, in stating that the bride was only twelve years old. Kästner and other biographers have been content to repeat the same assertion without any comment, notwithstanding its evident improbability. The origin of the blunder is to be found in Kepler's correspondence with Bernegger, to whom, speaking of his wife, he says." She has been educated for twelve years by the Lady of Stahrenberg." This is by no means a single instance of carelessness in Hantsch; Kästner has pointed out others of greater consequence. It was owing to this marriage, that Kepler took occasion to write his new method of gauging, for as he tells us in his own peculiar style "last November I brought home a new wife, and as the whole course of Danube was then covered with the produce of the Austrian vineyards, to be sold at a reasonable rate, I purchased a few casks, thinking it my duty as a good husband and a father of a family, to see that my household was well provided with drink." When the seller came to ascertain the quantity, Kepler objected to his method of gauging, for he allowed no difference, whatever might be the proportion of the bulging parts. The reflections to which this incident gave rise, terminated in the publication of the above-mentioned treatise, which claims a place among the earliest specimens of what is now called the modern analysis. In it he extended several properties of plane figures to segments of cones and cylinders, from the consideration that "these solids are incorporated circles," and, therefore, that those properties are true of the whole which belong to each component part. That the book might end as oddly as it began, Kepler concluded it with a parody of Catullus:

"Et cum pocula mille mensi erimus
Conturbabimus illa, ne sciamus."

His new residence at Linz was not long undisturbed. He quarrelled there, as he had done in the early part of his life at Gratz, with the Roman Catholic party, and was excommunicated. "Judge," says he to Peter Hoffman, "how far I can assist you, in a place where the priest and school-inspector have combined to brand me with the public stigma of heresy, because in every question I take that side which seems to me to be consonant with the word of God." The particular dogma which occasioned his excommunication, was connected with the doctrine of transubstantiation. He published his creed in a copy of Latin verses, preserved by his biographer Hantsch.

Before this occurrence, Kepler had been called to the diet at Ratisbon to give his opinion on the propriety of adopting the Gregorian reformation of the calendar, and he published a short essay, pointing out the respective convenience of doing so, or of altering the old Julian Calendar in some other manner. Notwithstanding the readiness of the diet to avail themselves of his talents for the settlement of a difficult question, the arrears of his salary were not paid much more regularly than they had been in Rodolph's time, and he was driven to provide himself with money by the publication of his almanac, of which necessity he heavily and justly complained. "In order to pay the expense of the Ephemeris for these two years, I have also written a vile prophesying almanac, which is scarcely more respectable than begging; unless it be because it saves the emperor's credit, who abandons me entirely; and with all his frequent and recent orders in council, would suffer me to perish with hunger." Kepler published this Ephemeris annually till 1620; ten years later he added those belonging to the years from 1620 to 1628.

In 1617 Kepler was invited into Italy, to succeed Magini as Professor of Mathematics at Bologna. The offer tempted him; but, after mature consideration, he rejected it, on grounds which he thus explained to Roffini:—" By birth and spirit I am a German, imbued with German principles, and bound by such family ties, that even if the emperor should consent, I could not, without the greatest difficulty, remove my dwelling-place from Germany into Italy. And although the glory of holding so distinguished a situation among the venerable professors of Bologna stimulates me, and there appears great likelihood of notably increasing my fortune, as well from the great concourse to the public lectures, as from private tuition; yet, on the other hand, that period of my life is past which was once excited by novelty, or which might promise itself a long enjoyment of these advantages. Besides, from a boy up to my present years, living a German among Germans, I am accustomed to a degree of freedom in my speech and manners, which, if persevered in on my removal to Bologna, seems likely to draw upon me, if not danger, at least notoriety, and might expose me to suspicion and party malice. Notwithstanding this answer, I have yet hopes that your most honourable invitation will be of service to me, and may make the imperial treasurer more ready than he has hitherto been to fulfil his master's intentions towards me. In that case I shall the sooner be able to publish the Rudolphine Tables and the Ephemerides, of which you had the scheme so many years back; and in this manner you and your advisers may have no reason to regret this invitation, though for the present it seems fruitless."

In 1619, the Emperor Matthias died, and was succeeded by Ferdinand III., who retained Kepler in the post he had filled under his two predecessors on the imperial throne. Kästner, in his "History of Mathematics," has corrected a gross error of Hantsch, in asserting that Kepler prognosticated Matthias's death. The letter to which Hantsch refers, in support of his statement, does indeed mention the emperor's death, but merely as a notorious event, for the purpose of recalling a date to the memory of his correspondent.

CHAPTER VII.

Kepler publishes his Harmonics—Account of his Astrological Opinions and Discovery of the Law of the Periods of the Planetary Revolutions—Sketch of Newton's proof of Kepler's Laws.

The "Cosmographical Mystery" was written, as has been already mentioned, when Kepler was only twenty-six, and the wildness of its theories might be considered as due merely to the vivacity of a young man; but as if purposely to shew that his maturer age did not renounce the creations of his youthful fancy, he reprinted the "Mystery" in 1619, nearly at the same time when he published his celebrated work on Harmonies; and the extravagance of the latter publication does not at all lose in comparison with its predecessor. It is dedicated to James I. of England, and divided into five books: "The first, Geometrical, on the origin and demonstration of the laws of the figures which produce harmonious proportions;—the second, Architectonical, on figurate geometry, and the congruence of plane and solid regular figures;—the third, properly Harmonic, on the derivation of musical proportions from figures, and on the nature and distinction of things relating to song, in opposition to the old theories;—the fourth, Metaphysical, Psychological, and Astrological, on the mental essence of harmonies, and of their kinds in the world, especially on the harmony of rays emanating on the earth from the heavenly bodies, and on their effect in nature, and on the sublunary and human soul;—the fifth, Astronomical and Metaphysical, on the very exquisite harmonies of the celestial motions, and the origin of the excentricities in harmonious proportions."

The two first books are almost strictly, as Kepler styles them, geometrical, relating in great measure to the inscription of regular polygons in a circle. The following passage is curious, presenting an analogous idea to that contained in one of the extracts already given from the Commentaries on Mars. "The heptagon, and all other polygons and stars beyond it, which have a prime number of sides, and all other figures derived from them, cannot be inscribed geometrically in a circle; although their sides have a necessary magnitude, it is equally a matter of necessity that we remain ignorant of it. This is a question of great importance, for on this account is it that the heptagon, and other figures of this kind, have not been employed by God in the adornment of the world, as the other intelligible figures are employed which have been already explained." Kepler then introduces the algebraical equation, on the solution of which this problem depends, and makes a remark which is curious at this period of the history of algebra—that the root of an equation which cannot be accurately found, may yet be found within any degree of approximation by an expert calculator. In conclusion he again remarks that "the side of the heptagon has no place among scientific existences, since its formal description is impossible, and therefore it cannot be known by the human mind, since the possibility of description precedes the possibility of knowledge; nor is it known even by the simple eternal act of an omniscient mind, because its nature belongs to things which cannot be known. And yet this scientific nonentity has some scientific properties, for if a heptagon were described in a circle, the proportion of its sides would have analogous proportions."

The third book is a treatise on music, in the confined and ordinary sense in which we now use that word, and apparently a sober and rational one, at least as nearly so as Kepler could be trusted to write on a subject so dangerous to his discretion. All the extravagance of the work seems reserved for the fourth book, the title of which already conveys some notion of the nature of its contents. In this book he has collected the substance of the astrological opinions scattered through his other works. We shall content ourselves with merely citing his own words, without any attempt to explain the difference between the astrology which he believed, and that which he contemptuously rejected. The distinctive line seems very finely drawn, and as both one and the other are now discarded by all who enjoy the full use of their reasoning powers, it is not of much consequence that it should be accurately traced.

It is to be observed, that he does not in this treatise modify or recant anything of his earlier opinions, but refers to the favourable judgment of his contemporary philosophers as a reason for embodying them in a regular form. "Since many very celebrated professors of philosophy and medicine are of opinion that I have created a new and most true philosophy, this tender plant, like all novelties, ought to be carefully nursed and cherished, so that it may strike root in the minds of philosophers, and not be choked by the excessive humours of vain sophistications, or washed away by the torrents of vulgar prejudices, or frozen by the chill of public neglect; and if I succeed in guarding it from these dangers, I have no fear that it will be crushed by the storms of calumny, or parched by the sun of sterling criticism."

One thing is very remarkable in Kepler's creed, that he whose candour is so indisputable in every other part of his conduct, professed to have been forced to adopt his astrological opinions from direct and positive observation.—" It is now more than twenty years since I began to maintain opinions like these on the predominant nature of the elements, which, adopting the common name, I call sublunary. I have been driven to this not by studying or admiring Plato, but singly and solely by observing seasons, and noting the aspects by which they are produced. I have seen the state of the atmosphere almost uniformly disturbed as often as the planets are in conjunction, or in the other configurations so celebrated among astrologers. I have noticed its tranquil state, either when there are none or few such aspects, or when they are transitory and of short duration. I have not formed an opinion on this matter without good grounds, like the common herd of prophesiers, who describe the operations of the stars as if they were a sort of deities, the lords of heaven and earth, and producing everything at their pleasure. They never trouble themselves to consider what means the stars have of working any effects among us on the earth, whilst they remain in the sky, and send down nothing to us which is obvious to the senses except rays of light. This is the principal source of the filthy astrological superstitions of that vulgar and childish race of dreamers, the prognosticators."

The real manner in which the configurations of the stars operate, according to Kepler, is as follows :—" Like one who listens to a sweet melodious song, and by the gladness of his countenance, by his voice, and by the beating of his hand or foot attuned to the music, gives token that he perceives and approves the harmony: just so does sublunary nature, with the notable and evident emotion of the bowels of the earth, bear like witness to the same feelings, especially at those times when the rays of the planets form harmonious configurations on the earth."—" I have been confirmed in this theory by that which might have deterred others ; I mean, by observing that the emotions do not agree nicely with the instants of the configurations; but the earth sometimes appears lazy and obstinate, and at another time (after important and long-continued configurations) she becomes exasperated, and gives way to her passion, even without the continuation of aspects. For in fact the earth is not an animal like a dog, ready at every nod ; but more like a bull, or an elephant, slow to become angry, and so much the more furious when incensed."

This singular doctrine must not be mistaken for one of Kepler's favourite allegories ; he actually and literally professed to believe that the earth was an enormous living animal; and he has enumerated, with a particularity of details into which we forbear to follow him, the analogies he recognized between its habits and those of men and other animals. A few samples of these may speak for the rest. " If any one who has climbed the peaks of the highest mountains throw a stone down their very deep clefts, a sound is heard from them; or if he throw it into one of the mountain lakes, which beyond doubt are bottomless, a storm will immediately arise, just as when you thrust a straw into the ear or nose of a ticklish animal, it shakes its head, or runs shuddering away. What so like breathing, especially of those fish who draw water into their mouths and spout it out again through their gills, as that wonderful tide! For although it is so regulated according to the course of the moon, that, in the preface to my ' Commentaries on Mars,' I have mentioned it as probable that the waters are attracted by the moon as iron is by the loadstone ; yet, if any one uphold that the earth regulates its breathing according to the motion of the sun and moon, as animals have daily and nightly alternations of sleep and waking, I shall not think his philosophy unworthy of being listened to; especially if any flexible parts should be discovered in the depths of the earth to supply the functions of lungs or gills."

From the next extract, we must leave the reader to learn as well as he may, how much Kepler did, and how much he did not believe on the subject of genethliac astrology.—" Hence it is that human spirits, at the time of celestial aspects, are particularly urged to complete the matters which they have in hand. What the goad is to the ox, what the spur or the rowel is to the horse, to the soldier the bell and trumpet, an animated speech to an audience, to a crowd of rustics a performance on the fife and bagpipes, that to all, and especially in the aggregate, is a heavenly configuration of suitable planets; so that every single one is excited in his thoughts and actions, and all become more ready to unite and associate their efforts. For instance, in war you may see that tumults, battles, fights, invasions, assaults, attacks, and panic fears, generally happen at the time of the aspects of Mars and Mercury, Mars and Jupiter, Mars and the Sun, Mars and Saturn, &c. In epidemic diseases, a greater number of persons are attacked at the times of the powerful aspects; they suffer more severely, or even die, owing to the failure of nature in her strife with the disease, which strife (and not the death) is occasioned by the aspect. It is not the sky which does all these things immediately, but the faculty of the vital soul, associating its operation with the celestial harmonies, is the principal agent in this so-called influence of the heavens. Indeed this word influence has so fascinated some philosophers that they prefer raving with the senseless vulgar, to learning the truth with me. This essential property is the principal foundation of that admirable genethliac art. For when anything begins to have its being when that is working harmonies, the sensible harmony of the rays of the planets has peculiar influence on it. This then is the cause why those who are born under a season of many aspects among the planets, generally turn out busy and industrious, whether they accustom themselves from childhood to amass wealth, or are born or chosen to direct public affairs, or finally, have given their attention to study. If any one think that I might be taken as an instance of this last class, I do not grudge him the knowledge of my nativity. I am not checked by the reproach of boastfulness, notwithstanding those who, by speech or conduct, condemn as folly all kinds of writing on this subject; the idiots, the half-learned, the inventors of titles and trappings, to throw dust in the eyes of the people, and those whom Picus calls the plebeian theologians: among the true lovers of wisdom, I easily clear myself of this imputation, by the advantage of my reader; for there is no one whose nativity or whose internal disposition and temper I can learn so well as I know my own. Well then, Jupiter ˋ nearest the nonagesimal had passed by four degrees the trine of Saturn; the Sun and Venus, in conjunction, were moving from the latter towards the former, nearly in sextiles with both: they were also removing from quadratures with Mars, to which Mercury was closely approaching: the moon drew near the trine of the same planet, close to the Bull's Eye, even in latitude. The 25th degree of Gemini was rising, and the 22d of Aquarius culminating. That there was this triple configuration on that day—namely, the sextile of Saturn and the Sun, the sextile of Mars and Jupiter, the quadrature of Mercury and Mars, is proved by the change of weather; for, after a frost of some days, that very day became warmer, there was a thaw and a fall of rain.*"

" I do not wish this single instance to be taken as a defence and proof of all the aphorisms of astrologers, nor do I attribute to the heavens the government of human affairs: what a vast interval still separates these philosophical observations from that folly or madness as it should rather be called. For, following up this example, I knew a lady†, born under nearly the same aspects, whose disposition, indeed, was exceedingly restless, but who not only makes no progress in literature (that is not strange in a woman), but troubles her whole family, and is the cause to herself of deplorable misery. What, in my case, assisted the aspects was—firstly, the fancy of my mother when pregnant with me, a great admirer of her mother-in-law, my grandmother, who had some knowledge of medicine, my grandfather's profession; a second cause is, that I

---

* This mode of verifying configurations, though something of the boldest, was by no means unusual. On a former occasion Kepler, wishing to cast the nativity of his friend Zehentmaier, and being unable to procure more accurate information than that he was born about three o'clock in the afternoon of the 21st of October, 1751, supplied the deficiency by a record of fevers and accidents at known periods of his life, from which he deduced a more exact horoscope.

† Kepler probably meant his own mother, whose horoscope he in many places declared to be nearly the same as his own.

L 2

was born a male, and not a female, for astrologers have sought in vain to distinguish sexes in the sky; thirdly, I derive from my mother a habit of body, more fit for study than other kinds of life: fourthly, my parents' fortune was not large, and there was no landed property to which I might succeed and become attached; fifthly, there were the schools, and the liberality of the magistracy towards such boys as were apt for learning. But now if I am to speak of the result of my studies, what I pray can I find in the sky, even remotely alluding to it. The learned confess that several not despicable branches of philosophy have been newly extricated or amended or brought to perfection by me: but here my constellations were, not Mercury from the east, in the angle of the seventh, and in quadratures with Mars, but Copernicus, but Tycho Brahe, without whose books of observations everything now set by me in the clearest light must have remained buried in darkness; not Saturn predominating Mercury, but my Lords the Emperors Rodolph and Matthias; not Capricorn, the house of Saturn, but Upper Austria, the home of the Emperor, and the ready and unexampled bounty of his nobles to my petition. Here is that corner, not the western one of the horoscope, but on the Earth, whither, by permission of my imperial master, I have betaken myself from a too uneasy court; and whence, during these years of my life, which now tends towards its setting, emanate these Harmonies, and the other matters on which I am engaged."

" However, it may be owing to Jupiter's ascendancy that I take greater delight in the application of geometry to physics, than in that abstract pursuit which partakes of the dryness of Saturn; and it is perhaps the gibbous moon, in the bright constellation of the Bull's forehead, which fills my mind with fantastic images."

The most remarkable thing contained in the 5th Book, is the announcement of the celebrated law connecting the mean distances of the planets with the periods of their revolution about the Sun. This law is expressed in mathematical language, by saying that the squares of the times vary as the cubes of the distances*. Kepler's rapture on detecting it was unbounded, as may be seen from the exulting rhapsody with which he announced it. " What I 'prophecied two-and-twenty years ago, as soon as I discovered the five solids among the heavenly orbits — what I firmly believed long before I had seen Ptolemy's ' Harmonics '—what I had promised my friends in the title of this book, which I named before I was sure of my discovery—what, sixteen years ago, I urged as a thing to be sought—that for which I joined Tycho Brahe, for which I settled in Prague, for which I have devoted the best part of my life to astronomical contemplations, at length I have brought to light, and have recognized its truth beyond my most sanguine expectations. Great as is the absolute nature of Harmonics with all its details, as set forth in my third book, it is all found among the celestial motions, not indeed in the manner which I imagined, (that is not the least part of my delight,) but in another very different, and yet most perfect and excellent. It is now eighteen months since I got the first glimpse of light, three months since the dawn, very few days since the unveiled sun, most admirable to gaze on, burst out upon me. Nothing holds me; I will indulge in my sacred fury; I will triumph over mankind by the honest confession, that I have stolen the golden vases of the Egyptians*, to build up a tabernacle for my God far away from the confines of Egypt. If you forgive me, I rejoice; if you are angry, I can bear it: the die is cast, the book is written; to be read either now or by posterity, I care not which: it may well wait a century for a reader, as God has waited six thousand years for an observer."

He has told, with his usual partienlarity, the manner and precise moment of the discovery. " Another part of my ' Cosmographical Mystery,' suspended twenty-two years ago, because it was then undetermined, is completed and introduced here, after I had discovered the true intervals of the orbits, by means of Brahe's observations, and had spent the continuous toil of a long time in investigating the true proportion of the periodic times to the orbits,

Sera quidem respexit inertem,
Respexit tamen, et longo post tempore venit.

If you would know the precise moment, the first idea came across me on the 8th March of this year, 1618; but chancing

---

* See Preliminary Treatise, p. 13.

* In allusion to the Harmonics of Ptolemy.

to make a mistake in the calculation, I rejected it as false. I returned again to it with new force on the 15th May, and it has dissipated the darkness of my mind by such an agreement between this idea and my seventeen years' labour on Brahe's observations, that at first I thought I must be dreaming, and had taken my result for granted in my first assumptions. But the fact is perfect, the fact is certain, that the proportion existing between the periodic times of any two planets is exactly the sesquiplicate proportion of the mean distances of the orbits."

There is high authority for not attempting over anxiously to understand the rest of the work. Delambre sums it up as follows:—"In the music of the celestial bodies it appears that Saturn and Jupiter take the bass, Mars the tenor, the Earth and Venus the counter-tenor, and Mercury the treble." If the patience of this indefatigable historian gave way, as he confesses, in the perusal, any further notice of it here may be well excused. Kepler became engaged, in consequence of this publication, in an angry controversy with the eccentric Robert Fludd, who was at least Kepler's match in wild extravagance and mysticism, if far inferior to him in genius. It is diverting to hear each reproaching the other with obscurity.

In the "Epitome of the Copernican Astronomy," which Kepler published about the same time, we find the manner in which he endeavoured to deduce the beautiful law of periodic times, from his principles of motion and radiation of whirling forces. This work is in fact a summary of all his astronomical opinions, drawn up in a popular style in the form of question and answer. We find there a singular argument against believing, as some did, that each planet is carried round by an angel, for in that case, says Kepler, "the orbits would be perfectly circular; but the elliptic form, which we find in them, rather smacks of the nature of the lever and material necessity."

The investigation of the relation between the periodic times and distances of the planets is introduced by a query whether or not they are to be considered heavy. The answer is given in the following terms:—"Although none of the celestial globes are heavy, in the sense in which we say on earth that a stone is heavy, nor light as fire is light with us, yet have they, by reason of their materiality, a natural inability to move from place to place: they have a natural inertness or quietude, in consequence of which they remain still in every situation where they are placed alone."

"*P.* Is it then the sun, which by its turning carries round the planets? How can the sun do this, having no hands to seize the planet at so great a distance, and force it round along with itself?—Its bodily virtue, sent forth in straight lines into the whole space of the world, serves instead of hands; and this virtue, being a corporeal species, turns with the body of the sun like a very rapid vortex, and travels over the whole of that space which it fills as quickly as the sun revolves in its very confined space round the centre.

"*P.* Explain what this virtue is, and belonging to what class of things?— As there are two bodies, the mover and the moved, so are there two powers by which the motion is obtained. The one is passive, and rather belonging to matter, namely, the resemblance of the body of the planet to the body of the sun in its corporeal form, and so that part of the planetary body is friendly, the opposite part hostile to the sun. The other power is active, and bearing more relation to form, namely, the body of the sun has a power of attracting the planet by its friendly part, of repelling it by the hostile part, and finally, of retaining it if it be placed so that neither the one nor the other be turned directly towards the sun.

"*P.* How can it be that the whole body of the planet should be like or cognate to the body of the sun, and yet part of the planet friendly, part hostile to the sun? —Just as when one magnet attracts another, the bodies are cognate; but attraction takes place only on one side, repulsion on the other.

"*P.* Whence, then, arises that difference of opposite parts in the same body? —In magnets the diversity arises from the situation of the parts with respect to the whole. In the heavens the matter is a little differently arranged, for the sun does not, like the magnet, possess only on one side, but in all the parts of its substance, this active and energetic faculty of attracting, repelling, or retaining the planet. So that it is probable that the centre of the solar body corresponds to one extremity or pole of the magnet, and its whole surface to the other pole.

"*P.* If this were so, all the planets

would be restored* in the same time with the sun?—True, if this were all: but it has been said already that, besides this carrying power of the sun, there is also in the planets a natural inertness to motion, which causes that, by reason of their material substance, they are inclined to remain each in its place. The carrying power of the sun, and the impotence or material inertness of the planet, are thus in opposition. Each shares the victory; the sun moves the planet from its place, although in some degree it escapes from the chains with which it was held by the sun, and so is taken hold of successively by every part of this circular virtue, or, as it may be called, solar circumference, namely, by the parts which follow those from which it has just extricated itself.

"*P.* But how does one planet extricate itself more than another from this violence—First, because the virtue emanating from the sun has the same degree of weakness at different distances, as the distances or the width of the circles described on these distances†. This is the principal reason. Secondly, the cause is partly in the greater or less inertness or resistance of the planetary globes, which reduces the proportions to onehalf; but of this more hereafter.

"*P.* How can it be that the virtue emanating from the sun becomes weaker at a greater distance? What is there to hurt or weaken it?—Because that virtue is corporeal, and partaking of quantity, which can be spread out and rarefied. Then, since there is as much virtue diffused in the vast orb of Saturn as is collected in the very narrow one of Mercury, it is very rare and therefore weak in Saturn's orbit, very dense and therefore powerful at Mercury.

"*P.* You said, in the beginning of this inquiry into motion, that the periodic times of the planets are exactly in the sesquiplicate proportion of their orbits or circles: pray what is the cause of this? —Four causes concur for lengthening the periodic time. First, the length of the path; secondly, the weight or quantity of matter to be carried; thirdly, the degree of strength of the moving virtue; fourthly, the bulk or space into which is spread out the matter to be moved. The circular paths of the planets are in the simple ratio of the distances; the weights or quantities of matter in different planets are in the subduplicate ratio of the same distances, as has been already proved; so that with every increase of distance, a planet has more matter, and therefore is moved more slowly, and accumulates more time in its revolution, requiring already as it did more time by reason of the length of the way. The third and fourth causes compensate each other in a comparison of different planets: the simple and subduplicate proportion compound the sesquiplicate proportion, which therefore is the ratio of the periodic times."

Three of the four suppositions here made by Kepler to explain the beautiful law he had detected, are now indisputably known to be false. Neither the weights nor the sizes of the different planets observe the proportions assigned by him, nor is the force by which they are retained in their orbits in any respect similar in its effects to those attributed by him to it. The wonder which might naturally be felt that he should nevertheless reach the desired conclusion, will be considerably abated on examining the mode in which he arrived at and satisfied himself of the truth of these three suppositions. It has been already mentioned that his notions on the existence of a whirling force emanating from the sun, and decreasing in energy at increased distances, are altogether inconsistent with all the experiments and observations we are able to collect. His reason for asserting that the sizes of the different planets are proportional to their distances from the sun, was simply because he chose to take for granted that either their solidities, surfaces, or diameters, must necessarily be in that proportion, and of the three, the solidities appeared to him least liable to objection. The last element of his precarious reasoning rested upon equally groundless assumptions. Taking as a principle, that where there is a number of different things they must be different in every respect, he declared that it was quite unreasonable to suppose all the planets of the same density. He thought it indisputable that they must be rarer as they were farther from the sun, "and yet not in the proportion of their distances, for thus we should sin against the law of variety in another way, and make the quantity of matter (according to what he had just said of their bulk) the same in

---

* This is a word borrowed from the Ptolemaic astronomy, according to which the sun and planets are hurried from their places by the daily motion of the *primum mobile*, and by their own peculiar motion seek to regain or be restored to their former places.

† In other parts of his works, Kepler assumes the diminution to be proportional to the circles themselves, not to the diameters.

all. But if we assume the ratio of the quantities of matter to be half that of the distances, we shall observe the best mean of all; for thus Saturn will be half as heavy again as Jupiter, and Jupiter half again as dense as Saturn. And the strongest argument of all is, that unless we assume this proportion of the densities, the law of the periodic times will not answer." This is the *proof* alluded to, and it is clear that by such reasoning any required result might be deduced from any given principles.

It may not be uninstructive to subjoin a sketch of the manner in which Newton established the same celebrated results, starting from principles of motion diametrically opposed to Kepler's, and it need scarcely be added, reasoning upon them in a manner not less different. For this purpose, a very few prefatory remarks will be found sufficient.

The different motions seen in nature are best analysed and classified by supposing that every body in motion, if left to itself, will continue to move forward at the same rate in a straight line, and by considering all the observed deviations from this manner of moving, as exceptions and disturbances occasioned by some external cause. To this supposed cause is generally given the name of Force, and it is said to be the first law of motion, that, unless acted on by some force, every body at rest remains at rest, and every body in motion proceeds uniformly in a straight line. Many employ this language, without perceiving that it involves a definition of force, on the admission of which, it is reduced to a truism. We see common instances of force in a blow, or a pull from the end of a string fastened to the body: we shall also have occasion presently to mention some forces where no visible connexion exists between the moving body and that towards which the motion takes place, and from which the force is said to proceed.

A second law of motion, founded upon experiment, is this: if a body have motion communicated to it in two directions, by one of which motions alone it would have passed through a given space in a given time, as for instance, through B C′ in one second, and by the other alone through any other space B c in the same time, it will, when both are given to it at the same instant, pass in the same time (in the present instance in one second) through B C the diagonal of the parallelogram of which B C′ and B c are sides.

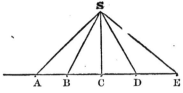

Let a body, acted upon by no force, be moving along the line A E ; that

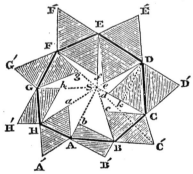

means, according to what has been said, let it pass over the equal straight lines A B, B C, C D, D E, &c., in equal times. If we take any point S not in the line A E, and join A S, B S, &c., the triangles A S B, B S C, &c. are also equal, having a common altitude and standing on equal bases, so that if a string were conceived reaching from S to the moving body (being lengthened or shortened in each position to suit its distance from S), this string, as the body moved along A E, would sweep over equal triangular areas in equal times.

Let us now examine how far these conclusions will be altered if the body from time to time is forced towards S. We will suppose it moving uniformly from A to B as before, no matter for the present how it got to A, or into the direction A B. If left to itself it would, in an equal time (say 1″) go through B C′ in the same straight line with and equal to A B. But just as it reaches B, and is beginning to move along B C′, let it be suddenly pulled towards S with a motion which, had it been at rest, would have carried it in the same time, 1″, through any other space B c. According to the second law of motion, its direction during this 1″, in consequence of the two motions combined, will be along B C, the diagonal of the parallelogram of which B C′, B c, are sides. In-this case, as this figure is drawn, B C, though passed in the same time, is longer than A B; that is to say, the body is moving quicker than at first. How is it with the triangular areas, supposed as before to be swept by a string constantly stretched between S and the body? It will soon be seen that these still remain equal, notwithstanding the change of direction, and increased swiftness. For since .C C′ is parallel to B c, the triangles S C B, S C′ B are equal, being on the same base S B, and between the same parallels S B, C C′, and S C′B is equal to·S B A as before, therefore S C B, S B A are equal. The body is now moving uniformly (though quicker than along A B) along B C. As before, it would in a time equal to the time of passing along B C, go through an equal space C D′ in the same straight line. But if at C it has a second pull towards S, strong enough to carry it to d in the same time, its direction will change a second time to C D, the diagonal of the parallelogram, whose sides are C D′, C d; and the circumstances being exactly similar to those at the first pull, it is shewn in the same manner that the triangular area S D C = S C B = S B A.

Thus it appears, that in consequence of these intermitting pulls towards S, the body may be moving round, sometimes faster, sometimes slower, but that the triangles formed by any of the straight portions of its path (which are all described in equal times), and the lines joining S to the ends of that portion, are all equal. The path it will take depends of course, in other respects, upon the frequency and strength of the different pulls, and it might happen, if they were duly proportionate, that when at H, and moving off in the direction H A′, the pull H a might be such as just to carry the body back to A, the point from which it started, and with such a motion, that after one pull more, A b, at A, it might move along A B as it did at first. If this were so, the body would continue to move round in the same polygonal path, alternately approaching and receding from S, as long as the same pulls were repeated in the same order, and at the same intervals.

It seems almost unnecessary to remark, that the same equality which subsists between any two of these triangular areas subsists also between an equal number of them, from whatever part of the path taken ; so that, for instance, the four paths A B, B C, C D, D E, corresponding to the four areas A S B, B S C, C S D, D S E, that is, to the area A B C D E S, are passed in the same time as the four E F, F G, G H, H A, corresponding to the equal area E F G H A S. Hence it may be seen, if the whole time of revolution from A round to A again be called a year, that in half a year the body will have got to E, which in the present figure is more than half-way round, and so of any other periods.

The more frequently the pulls are supposed to recur, the more frequently will the body change its direction; and if the pull were supposed constantly exerted in the direction towards S, the body would move in a curve round S, for no three successive positions of it could be in a straight line. Those who are not familiar with the methods of measuring curvilinear spaces must here be contented to observe, that the law holds, however close the pulls are brought together, and however closely the polygon is consequently made to resemble a curve: they may, if they please, consider the minute portions into which the curve is so divided, as differing insensibly from little rectilinear triangles, any equal number of which, according to what has been said above, wherever taken in the curve, would be swept in equal times. The theorem admits, in this case also, a rigorous proof; but it is not easy to make it entirely satisfactory, without entering into explanations which would detain us too long from our principal subject.

The proportion in which the pull is strong or weak at different distances from the central spot, is called " *the law of the central or centripetal force*," and it may be observed, that after assuming the laws of motion, our investigations cease to have anything hypothetical or experimental in them ; and that if we wish, according to these principles of motion, to determine the law of force necessary to make a body move in a curve of any required form, or conversely to discover the form of the curve described, in consequence of any assumed law of force, the inquiry is purely geometrical, depending upon the nature and properties of geometrical quantities only. This distinction between what is hypothetical, and what necessary truth, ought never to be lost sight of.

As the object of the present treatise is not to teach geometry, we shall describe, in very general terms, the manner in which Newton, who was the first who systematically extended the laws of motion to the heavenly bodies, identified their results with the two remaining laws of Kepler. His " Principles of Natural Philosophy" contain general propositions with regard to any law of centripetal force, but that which he supposed to be the true one in our system, is expressed in mathematical language, by saying that the centripetal force varies inversely as the square of the distance, which means, that if the force at any distance be taken for the unit of force, at half that distance, it is two times twice, or four times as strong; at one-third the distance, three times thrice, or nine times as strong, and so for other distances. He shewed the probability of this law in the first instance by comparing the motion of the moon with that of heavy bodies at the surface of the earth. Taking L P to represent part of the moon's orbit described in one minute, the line P M between the orbit and the tangent at L would shew the space through which the central force at the earth (assuming the above principles of motion to be correct) would draw the moon. From the known distance and motion of the moon, this line P M is found to be about sixteen feet. The distance of the moon is about sixty times the radius of the earth, and therefore if the law of the central force in this instance were such as has been supposed, the force at the earth's surface would be 60 times 60, or 3600 times stronger, and at the earth's surface, the central force would make a body fall through 3600 times 16 feet in one minute. Galileo had already taught that the spaces through which a body would be made to fall, by the constant action of the same unvarying force, would be proportional to the squares of the times during which the force was exerted, and therefore according to these laws, a body at the earth's surface ought (since there are sixty seconds in a minute) to fall through 16 feet in one second, which was precisely the space previously established by numerous experiments.

With this confirmation of the supposition, Newton proceeded to the purely geometrical calculation of the law of centripetal * force necessary to make a moving body describe an ellipse round its focus, which Kepler's observations had established to be the form of the orbits of the planets round the sun. The result of the inquiry shewed that this curve required the same law of the force, varying inversely as the square of the distance, which therefore of course received additional confirmation. His method of doing this may, perhaps, be understood by referring to the last figure but one, in which C d, for instance, representing the space fallen from any point C towards S, in a given time, and the area C S D being proportional to the corresponding time, the space through which the body would have fallen at C in any other time (which would be greater, by Galileo's law, in proportion to the squares of the times), might be represented by a quantity varying directly as C d, and inversely in the duplicate proportion of the triangular area C S D, that is to say, proportional to

$$\frac{C\,d}{(S\,C \times D\,k)^2},$$ if D k be drawn from D

perpendicular on S C. If this polygon represent an ellipse, so that C D represents a small arc of the curve, of which S is the focus, it is found by the nature of that curve, that $\frac{C\,d}{(D\,k)^2}$ is the same at all points of the curve, so that the law of variation of the force in the same ellipse is represented solely by $\frac{1}{(S\,C)^2}$. If C d, &c. are drawn so that $\frac{C\,d}{(D\,k)^2}$ is not the same at every point, the curve ceases to be an ellipse whose focus is at S, as Newton has shewn in the same work. The line to which $\frac{(D\,k)^2}{C\,d}$ is found to be equal, is one drawn through the focus at right angles to the longest axis of the ellipse till it meets the curve;—this line is called the *latus rectum*, and is a third proportional to the two principal axes.

Kepler's third law follows as an immediate consequence of this determination; for, according to what has been already shown, the time of revolution round the whole ellipse, or, as it is com-

---

there is a point to which the name of centre is given, on account of peculiar properties belonging to it: but the term " centripetal force" always refers to the place towards which the force is directed, whether or not situated in the centre of the curve.

* In many curves, as in the circle and ellipse,

monly called, the periodic time, bears the same ratio to the unit of time as the whole area of the ellipse does to the area described in that unit. The area of the whole ellipse is proportional in different ellipses to the rectangle contained by the two principal axes, and the area described in an unit of time is proportional to S C × D k, that is to say, is in the subduplicate ratio of S C^2 × D k^2, or $\dfrac{D k^2}{C d}$, when the force varies inversely as the square of the distance S C ; and in the ellipse, as we have said already, this is equal to a third proportional to the principal axes; consequently the periodic times in different ellipses, which are proportional to the whole areas of the ellipses directly, and the areas described in the unit of time inversely, are in the compound ratio of the rectangle of the axes directly, and subduplicatly as a third proportional to the axes inversely; that is to say, the squares of these times are proportional to the cubes of the longest axes, which is Kepler's law.

CHAPTER VIII.

The Epitome prohibited at Rome—Logarithmic Tables—Trial of Catharine Kepler—Kepler invited to England—Rudolphine Tables—Death— Conclusion.

KEPLER'S "Epitome," almost immediately on its appearance, enjoyed the honour of being placed by the side of the work of Copernicus, on the list of books prohibited by the congregation of the Index at Rome. He was considerably alarmed on receiving this intelligence, anticipating that it might occasion difficulties in publishing his future writings. His words to Remus, who had communicated the news to him, are as follows:—
" I learn from your letter, for the first time, that my book is prohibited at Rome and Florence. I particularly beg of you, to send me the exact words of the censure, and that you will inform me whether that censure would be a snare for the author, if he were caught in Italy, or whether, if taken, he would be enjoined a recantation. It is also of consequence for me to know whether there is any chance of the same censure being extended into Austria. For if this be so, not only shall I never again find a printer there, but also the copies which the bookseller has left in Austria at my desire will be endangered, and the ultimate loss will fall upon me. It will amount to giving me to understand, that I must cease to profess Astronomy, after I have grown old in the belief of these opinions, having been hitherto gainsayed by no one,—and, in short, I must give up Austria itself, if room is no longer to be left in it for philosophical liberty." He was, however, tranquillized, in a great degree, by the reply of his friend, who told him that "the book is only prohibited as contrary to the decree pronounced by the holy office two years ago. This has been partly occasioned by a Neapolitan monk (Foscarini), who was spreading these notions by publishing them in Italian, whence were arising dangerous cousequences and opinions: and besides, Galileo was at the same time pleading his cause at Rome with too much violence. Copernicus has been corrected in the same manner for some lines, at least in the beginning of his first book. But by obtaining a permission, they may be read (and, as I suppose, this " Epitome" also) by the learned and skilful in this science, both at Rome and throughout all Italy. There is therefore no ground for your alarm, either in Italy or Austria; only keep yourself within bounds, and put a guard upon your own passions.".

We shall not dwell upon Kepler's different works on comets, beyond mentioning that they were divided, on the plan of many of his other publications, into three parts, Astronomical, Physical, and Astrological. He maintained that comets move in straight lines, with a varying degree of velocity. Later theories have shewn that they obey the same laws of motion as the planets, differing from them only in the extreme excentricity of their orbits. In the second book, which contains the Physiology of Comets, there is a passing remark that comets come out from the remotest parts of ether, as whales and monsters from the depth of the sea; and the suggestion is thrown out that perhaps comets are something of the nature of silkwoms, and are wasted and consumed in spinning their own tails.

Among his other laborious employments, Kepler yet found time to calculate tables of logarithms, he having been one of the first in Germany to appreciate the full importance of the facilities they afford to the numerical calculator. In 1618 he wrote to his friend Schickhard: " There is a Scottish Baron (whose name has escaped my memory), who has made a famous contrivance, by which all need of multiplication and division is supplied by mere addition and subtraction; and he does it without sines. But even he wants a table of tangents *, and the variety, frequency, and difficulty of the additions and subtractions, in some cases, is greater than the labour of multiplying and dividing."

Kepler dedicated his "Ephemeris" for 1620 to the author of this celebrated invention, Baron Napier, of Merchistoun; and in 1624, published what he called "Chilias Logarithmorum," containing the Napierian logarithms of the quotients of 100,000 divided by the first ten numbers, then proceeding by the quotients of every ten to 100, and by hundreds to 100,000. In the supplement published the following year, is a curious notice of the manner in which this subtle contrivance was at first received: "In the year 1621, when I had gone into Upper Austria, and had conferred everywhere with those skilled in mathematics, on the subject of Napier's logarithms, I found that those whose prudence had increased, and whose readiness had diminished, through age, were hesitating whether to adopt this new sort of numbers, instead of a table of sines; because they said it was disgraceful to a professor of mathematics to exult like a child at some compendious method of working, and meanwhile to admit a form of calculation, resting on no legitimate proof, and which at some time might entangle us in error, when we least feared it. They complained that Napier's demonstration rested on a fiction of geometrical motion, too loose and slippery for a sound method of reasonable demonstration to be founded on it†. "This led me forthwith to conceive the germ of a legitimate demonstration, which during that same winter I attempted, without reference to lines or motion, or flow, or any other which I may call sensible quality."

"Now to answer the question; what is the use of logarithms? Exactly what ten years ago was announced by their author, Napier, and which may be told in these words.—Wheresoever in common arithmetic, and in the Rule of Three, come two numbers to be multiplied together, there the sum of the logarithms is to be taken; where one number is to be divided by another, the difference; and the number corresponding to this sum or difference, as the case may be, will be the required product or quotient. This, I say, is the use of logarithms. But in the same work in which I gave the demonstration of the principles, I could not satisfy the unfledged arithmetical chickens, greedy of facilities, and gaping with their beaks wide open, at the mention of this use, as if to bolt down every particular gobbet, till they are crammed with my precepticles."

The year 1622 was marked by the catastrophe of a singular adventure which befell Kepler's mother, Catharine, then nearly seventy years old, and by which he had been greatly harassed and annoyed during several years. From her youth she had been noted for a rude and passionate temper, which on the present occasion involved her in serious difficulties. One of her female acquaintance, whose manner of life had been by no means unblemished, was attacked after a miscarriage by violent headaches, and Catharine, who had often taken occasion to sneer at her notorious reputation, was accused with having produced these consequences, by the administration of poisonous potions. She repelled the charge with violence, and instituted an action of scandal against this person, but was unlucky (according to Kepler's statement) in the choice of a young doctor, whom she employed as her advocate. Considering the suit to be very instructive, he delayed its termination during five years, until the judge before whom it was tried was displaced. He was succeeded by another, already indisposed against Catharine Kepler, who on some occasion had taunted him with his sudden accession to wealth from a very inferior situation. Her opponent, aware of this advantage, turned the ta-

---

* The meaning of this passage is not very clear: Kepler evidently had seen and used logarithms at the time of writing this letter; yet there is nothing in the method to justify this expression,—"*At tamen opus est ipsi Tangentium canone.*"

† This was the objection originally made to Newton's "Fluxions," and in fact, Napier's idea of logarithms is identical with that method of conceiving quantities. This may be seen at once from a few of his definitions,

1 Def. A line is said to increase uniformly, when the point by which it is described passes through equal intervals, in equal times.

2 Def. A line is said to diminish to a shorter one proportionally, when the point passing along it cuts off in equal times segments proportional to the remainder.

6 Def. The logarithm of any sine is the number most nearly denoting the line, which has increased uniformly, whilst the radius has diminished to that sine proportionally, the initial velocity being the same in both motions. (Mirifici logarithmorum canonis descriptio, Edinburgi 1614.)

. This last definition contains what we should now call the differential equation between a number and the logarithm of its reciprocal.

bles on her, and in her turn became the accuser. The end of the matter was, that in July, 1620, Catharine was imprisoned, and condemned to the torture. Kepler was then at Linz, but as soon as he learned his mother's danger, hurried to the scene of trial. He found the charges against her supported only by evidence which never could have been listened to, if her own intemperate conduct had not given advantage to her adversaries. He arrived in time to save her from the question, but she was not finally acquitted and released from prison till November in the following year. Kepler then returned to Linz, leaving behind him his mother, whose spirit seemed in no degree broken by the unexpected turn in the course of her litigation. She immediately commenced a new action for costs and damages against the same antagonist, but this was stopped by her death, in April 1622, in her seventy-fifth year.

In 1620 Kepler was visited by Sir Henry Wotton, the English ambassador at Venice, who finding him, as indeed he might have been found at every period of his life, oppressed by pecuniary difficulties, urged him to go over to England, where he assured him of a welcome and honourable reception; but Kepler could not resolve upon the proposed journey, although in his letters he often returned to the consideration of it. In one of them, dated a year later, he says, "The fires of civil war are raging in Germany—they who are opposed to the honour of the empire are getting the upper hand—everything in my neighbourhood seems abandoned to flame and destruction. Shall I then cross the sea, whither Wotton invites me? I, a German? a lover of firm land? who dread the confinement of an island? who presage its dangers, and must drag along with me my little wife and flock of children? Besides my son Louis, now thirteen years old, I have a marriageable daughter, a two-year old son by my second marriage, an infant daughter, and its mother but just recovering from her confinement." Six years later, be says again,—"As soon as the Rudolphine Tables are published, my desire will be to find a place where I can lecture on them to a considerable assembly; if possible, in Germany; if not, why then in Italy, France, the Netherlands, or England, provided the salary is adequate for a traveller."

In the same year in which he received this invitation an affront was put upon Kepler by his early patrons, the States of Styria, who ordered that all the copies of his "Calendar," for 1624, to be publicly burnt. Kepler declares that the reason of this was, that he had given precedence in the title-page to the States of Upper Ens, in whose service he then was, above Styria. As this happened during his absence in Wirtemberg, it was immediately coupled by rumour with his hasty departure from Linz: it was said that he had incurred the Emperor's displeasure, and that a large sum was set upon his head. At this period Matthias had been succeeded by Ferdinand III., who still continued to Kepler his barren title of imperial mathematician.

In 1624 Kepler went to Vienna, in the hopes of getting money to complete the Rudolphine Tables, but was obliged to be satisfied with the sum of 6000 florins and with recommendatory letters to the States of Suabia, from whom he also collected some money due to the emperor. On his return he revisited the University of Tubingen, where he found his old preceptor, Mästlin, still alive, but almost worn out with old age. Mästlin had well deserved the regard. Kepler always appears to have entertained for him; he had treated him with great liberality whilst at the University, where he refused to receive any remuneration for his instruction. Kepler took every opportunity of shewing his gratitude; even whilst he was struggling with poverty he contrived to send his old master a handsome silver cup, in acknowledging the receipt of which Mästlin says,—"Your mother had taken it into her head that you owed me two hundred florins, and had brought fifteen florins and a chandelier towards reducing the debt, which I advised her to send to you. I asked her to stay to dinner, which she refused: however, we handselled your cup, as you know she is of a thirsty temperament."

The publication of the Rudolphine Tables, which Kepler always had so much at heart, was again delayed, notwithstanding the recent grant, by the disturbances arising out of the two parties into which the Reformation had divided the whole of Germany. Kepler's library was sealed up by desire of the Jesuits, and nothing but his connexion with the Imperial Court secured to him his own personal indemnity. Then followed a popular insurrection, and the peasantry blockaded Linz, so that it was not until 1627 that these celebrated tables finally made their appearance, the earliest calculated on the supposition that the planets move in elliptic orbits. Ptolemy's tables had been succeeded by the "Alphonsine," so called from Alphonso, King of Castile, who, in the thirteenth century, was an enlightened patron of astronomy. After the discoveries of Copernicus, these again made way for the Prussian, or Prutenic tables, calculated by his pupils Reinhold and Rheticus. These remained in use till the observations of Tycho Brahe showed their insufficiency, and Kepler's new theories enabled him to improve upon them. The necessary types for these tables were cast at Kepler's own expense. They are divided into four parts, the first and third containing a variety of logarithmic and other tables, for the purpose of facilitating astronomical calculations. In the second are tables of the elements of the sun, moon, and planets. The fourth gives the places of 1000 stars as determined by Tycho, and also at the end his table of refractions, which appears to have been different for the sun, moon, and stars. Tycho Brahe assumed the horizontal refraction of the sun to be 7' 30", of the moon 8', and of the other stars 3'. He considered all refraction of the atmosphere to be insensible above 45° of altitude, and even at half that altitude in the case of the fixed stars. A more detailed account of these tables is here obviously unsuitable: it will be sufficient to say merely, that if Kepler had done nothing in the course of his whole life but construct these, he would have well earned the title of a most useful and indefatigable calculator.

Some copies of these tables have prefixed to them a very remarkable map, divided by hour lines, the object of which is thus explained:—

"The use of this nautical map is, that if at a given hour the place of the moon is known by its edge being observed to touch any known star, or the edges of the sun, or the shadow of the earth; and if that place shall (if necessary) be reduced from apparent to real by clearing it of parallax; and if the hour at Uraniburg be computed by the Rudolphine tables, when the moon occupied that true place, the difference will show the observer's meridian, whether the picture of the shores be accurate or not, for by this means it may come to be corrected."

This is probably one of the earliest announcements of the method of determining longitudes by occultations; the imperfect theory of the moon long remained a principal obstacle to its introduction in practice. Another interesting passage connected with the same object may be introduced here. In a letter to his friend Cruger, dated in 1616, Kepler says: "You propose a method of observing the distances of places by sundials and automata. It is good, but needs a very accurate practice, and confidence in those who have the care of the clocks. Let there be only one clock, and let it be transported; and in both places let meridian lines be drawn with which the clock may be compared when brought. The only doubt remaining is, whether a greater error is likely from the unequal tension in the automaton, and from its motion, which varies with the state of the air, or from actually measuring the distances. For if we trust the latter, we can easily determine the longitudes by observing the differences of the height of the pole."

In an Appendix to the Rudolphine Tables, or, as Kepler calls it, "an alms doled out to the nativity casters," he has shown how they may use his tables for their astrological predictions. Everything in his hands became an allegory; and on this occasion he says, —"Astronomy is the daughter of Astrology, and this modern Astrology, again, is the daughter of Astronomy, bearing something of the lineaments of her grandmother; and, as I have already said, this foolish daughter, Astrology, supports her wise but needy mother, Astronomy, from the profits of a profession not generally considered creditable."

Soon after the publication of these tables, the Grand Duke of Tuscany sent him a golden chain; and if we remember the high credit in which Galileo stood at this time in Florence, it does not seem too much to attribute this honourable mark of approbation to his representation of the value of Kepler's services to astronomy. This was soon followed by a new and final change in his fortunes. He received permission from the emperor to attach himself to the celebrated Duke of Friedland, Albert Wallenstein, one of the most remarkable men in the history of that time.

Wallenstein was a firm believer in astrology, and the reception Kepler experienced by him was probably due, in great measure, to his reputation in that art. However that may be, Kepler found in him a more munificent patron than any one of his three emperors; but he was not destined long to enjoy the appearance of better fortune. Almost the last work which he published was a commentary on the letter addressed, by the missionary Terrentio, from China, to the Jesuits at Ingolstadt. The object of this communication was to obtain from Europe means for carrying into effect a projected scheme for improving the Chinese calendar. In this essay Kepler maintains the opinion, which has been discussed with so much warmth in more modern times, that the pretended ancient observations of the Chinese were obtained by computing them backwards from a much more recent date. Wallenstein furnished him with an assistant for his calculations, and with a printing press; and through his influence nominated him to the professorship in the University of Rostoch, in the Duchy of Mecklenburg. His claims on the imperial treasury, which amounted at this time to 8000 crowns, and which Ferdinand would gladly have transferred to the charge of Wallenstein, still remained unsatisfied. Kepler made a last attempt to obtain them at Ratisbon, where the imperial meeting was held, but without success. The fatigue and vexation occasioned by his fruitless journey brought on a fever, which unexpectedly put an end to his life, in the early part of November, 1630, in his fifty-ninth year. His old master, Mästlin, survived him for about a year, dying at the age of eighty-one.

Kepler left behind him two children by his first wife, Susanna and Louis; and three sons and two daughters, Sebald, Cordelia, Friedman, Hildebert, and Anna Maria, by his widow. Susanna married, a few months before her father's death, a physician named Jacob Bartsch, the same who latterly assisted Kepler in preparing his "Ephemeris." He died very shortly after Kepler himself. Louis studied medicine, and died in 1663, whilst practising as a physician at Konigsberg. The other children died young.

Upon Kepler's death the Duke of Friedland caused an inventory to be taken of his effects, when it appeared that near 24,000 florins were due to him, chiefly on account of his salary from the emperor. His daughter Susanna, Bartsch's widow, managed to obtain a part of these arrears by refusing to give up Tycho Brahe's observations till her claims were satisfied. The widow and younger children were left in very straightened circumstances, which induced Louis, Kepler's eldest son, to print, for their relief, one of his father's works, which had been left by him unpublished. It was not without much reluctance, in consequence of a superstitious feeling which he did not attempt to conceal or deny. Kepler himself, and his son-in-law, Bartsch, had been employed in preparing it for publication at the time of their respective deaths; and Louis confessed that he did not approach the task without apprehension that he was incurring some risk of a similar fate. This little rhapsody is entitled a "Dream on Lunar Astronomy;" and was intended to illustrate the appearances which would present themselves to an astronomer living upon the moon.

The narrative in the dream is put into the mouth of a personage, named Duracoto, the son of an Icelandic enchantress, of the name of Fiolxhildis. Kepler tells us that he chose the last name from an old map of Europe in his house, in which Iceland was called Fiolx: Duracoto seemed to him analogous to the names he found in the history of Scotland, the neighbouring country. Fiolxhildis was in the habit of selling winds to mariners, and used to collect herbs to use in her incantations on the sides of Mount Hecla, on the Eve of St. John. Duracoto cut open one of his mother's bags, in punishment of which she sold him to some traders, who brought him to Denmark, where he became acquainted with Tycho Brahe. On his return to Iceland, Fiolxhildis received him kindly, and was delighted with the progress he had made in astronomy. She then informed him of the existence of certain spirits, or demons, from whom, although no traveller herself, she acquired a knowledge of other countries, and especially of a very remarkable country, called Livania. Duracoto requesting further information, the necessary ceremonies were performed for invoking the demon; Duracoto and his mother enveloped their heads in their clothing, and presently "the screaking of a harsh dissonant voice began to speak in the Icelandic tongue." The island of Livania is situated in the depths of ether, at the distance of about 250000 miles; the road thence or thither is very seldom open, and even when it is passable, mankind find the journey a most difficult and dangerous one. The demon describes the method employed by his fellow spirits to convey such travellers as are thought fit for the undertaking: "We bring no sedentary people into our company, no corpulent or delicate persons; but we pick out those who waste their life in the continual use of post-horses, or who sail frequently to the Indies; who are accustomed to live upon biscuit, garlic, dried fish, and such abominable feeding. Those withered old hags are exactly fit for us, of whom the story is familiar that they travel immense distances by night on goats, and forks, and old petticoats. The Germans do not suit us at all; but we do not reject the dry Spaniards." This extract will probably be sufficient to show the style of the work. The inhabitants of Livania are represented to be divided into two classes, the Privolvans and Subvolvans, by whom are meant those supposed to live in the hemisphere facing the earth, which is called the Volva, and those on the opposite half of the moon: but there is nothing very striking in the account given of the various phenomena as respects these two classes. In some notes which were added some time after the book was first written, are some odd insights into Kepler's method of composing. Fiolxhildis had been made to invoke the dæmon with twenty-one characters; Kepler declares, in a note, that he cannot remember why he fixed on this number, "except because that is the number of letters in *Astronomia Copernicana*, or because there are twenty-one combinations of the planets, two together, or because there are twenty-one different throws upon two dice." The dream is abruptly terminated by a storm, in which, says Kepler, " I suddenly waked; the Demon, Duracoto, and Fiolxhildis were gone, and instead of their covered heads, I found myself rolled up among the blankets."

Besides this trifle, Kepler left behind him a vast mass of unpublished writings, which came at last into the hands of his biographer, Hantsch. In 1714, Hantsch issued a prospectus for publishing them by subscription, in twenty-two folio volumes. The plan met no encouragement, and nothing was published but a single folio volume of letters to and from Kepler, which seem to have furnished the principal materials for the memoir prefixed to them. After various unavailing attempts to interest different learned bodies in their appearance, the manuscripts were purchased for the library at St. Petersburg, where Euler, Lexell, and Kraft, undertook to examine them, and select the most interesting parts for publication. The result of this examination does not appear.

Kepler's body was buried in St. Peter's churchyard at Ratisbon, and a simple inscription was placed on his tombstone. This appears to have been destroyed not long after, in the course of the wars which still desolated the country. In 1786, a proposal was made to erect a marble monument to his memory, but nothing was done. Kästner, on whose authority it is mentioned, says upon this, rather bitterly, that it matters little whether or not Germany, having almost refused him bread during his life, should, a century and a half after his death, offer him a stone.

Delambre mentions, in his History of Astronomy, that this design was resumed in 1803 by the Prince Bishop of Constance, and that a monument has been erected in the Botanical Garden at Ratisbon, near the place of his interment. It is built in the form of a temple, surmounted by a sphere; in the centre is placed a bust of Kepler, in Carrara marble. Delambre does not mention the original of the bust; but says it is not unlike the figure engraved in the frontispiece of the Rudolphine Tables. That frontispiece consists of a portico of ten pillars, supporting a cupola covered with astronomical emblems. Copernicus, Tycho Brahe, Ptolemy, Hipparchus, and other astronomers, are seen among them. In one of the compartments of the common pedestal is a plan of the observatory at Uraniburg; in another, a printing press; in a third is the figure of a man, meant for Kepler, seated at a table. He is identified by the titles of his works, which are round him; but the whole is so small as to convey very little idea of his figure or countenance. The only portrait known of Kepler was given by him to his assistant Gringallet, who presented it to Bernegger; and it was placed by the latter in the library at Strasburg. Hantsch had a copy taken for the purpose of engraving it, but died before it was completed. A portrait of Kepler is engraved in the seventh part of Boissard's Bibliotheca Chalcographica. It is not known whence this was taken, but it may, perhaps, be a copy of that which was engraved by desire of Bernegger in 1620. The likeness is said not to have been well preserved. " His heart and genius," says Kästner, " are faithfully depicted in his writings; and that may console us, if we cannot entirely trust his portrait." In the preceding pages, it has been endeavoured to select such passages from his writings as might throw the greatest light on his character, with a subordinate reference only to the importance of the subjects treated. In conclusion, it may be well to support the opinion which has been ventured on the real nature of his triumphs, and on the danger of attempting to follow his method in the pursuit of truth, by the judgment pronounced by Delambre, as well on his failures as on his success. " Considering these matters in another point of view, it is not impossible to convince ourselves that Kepler may have been always the same. Ardent, restless, burning to distinguish himself by his discoveries, he attempted everything; and having once obtained a glimpse of one, no labour was too hard for him in following or verifying it. All his attempts had not the same success, and, in fact, that was impossible. Those which have failed seem to us only fanciful; those which have been more fortunate appear sublime. When in search of that which really existed, he has sometimes found it; when he devoted himself to the pursuit of a chimera, he could not but fail; but even there he unfolded the same qualities, and that obstinate perseverance that must triumph over all difficulties but those which are insurmountable*."

---

* Histoire de l'Astronomie Moderne, Paris, 1821.

List of Kepler's published Works.

Ein Calender	Gratz, 1594
Prodromus Dissertat. Cosmograph.	Tubingæ, 1596, 4to.
De fundamentis Astrologiæ	Pragæ, 1602, 4to.
Paralipomena ad Vitellionem	Francofurti, 1604, 4to.
Epistola de Solis deliquio	1605
De stellâ novâ	Pragæ, 1606, 4to.
Vom Kometen	Halle, 1608, 4to.
Antwort an Röslin	Pragæ, 1609, 4to.
Astronomia Nova	Pragæ, 1609, fol.
Tertius interveniens	Frankfurt, 1610, 4to.
Dissertatio cum Nuncio Sidereo	Francofurti, 1610, 4to.
Strena, seu De nive sexangulâ	Frankfurt, 1611, 4to.
Dioptrica	Francofurti, 1611, 4to.
Vom Geburts Jahre des Heylandes	Strasburg, 1613, 4to.
Respons. ad epist S. Calvisii	Francofurti, 1614, 4to.
Eclogæ Chronicæ	Frankfurt, 1615, 4to.
Nova Stereometria	Lincii, 1615, 4to.
Ephemerides 1617—1620	Lincii, 1616, 4to.
Epitomes Astron. Copern. Libri i. ii. iii.	Lentiis, 1618, 8vo.
De Cometis	Aug. Vindelic. 1619, 4to.
Harmonice Mundi	Lincii, 1619, fol.
Kanones Pueriles	Ulmæ, 1620
Epitomes Astron. Copern. Liber iv.	Lentiis, 1622, 8vo.
Epitomes Astron. Copern. Libri v. vi. vii.	Francofurti, 1622, 8vo.
Discurs von der grossen Conjunction	Linz. 1623, 4to.
Chilias Logarithmorum	Marpurgi, 1624, fol.
Supplementum	Lentiis, 1625, 4to.
Hyperaspistes	Francofurti, 1625, 8vo.
Tabulæ Rudolphinæ	Ulmæ, 1627, fol.
Resp. ad epist. J. Bartschii	Sagani, 1629, 4to.
De anni 1631 phænomenis	Lipsæ, 1629, 4to.
Terrentii epistolium cum commentatiunculâ	Sagani, 1630, 4to.
Ephemerides	Sagani, 1630, 4to.
Somnium	Francofurti, 1634, 4to.
Tabulæ manuales	Argentorati, 1700, 12mo.

12/60

RETURN	CIRCULATION DEPARTMENT	
TO ➡	202 Main Library	
LOAN PERIOD 1 **HOME USE**	2	3
4	5	6

ALL BOOKS MAY BE RECALLED AFTER 7 DAYS
1-month loans may be renewed by calling 642-3405
6-month loans may be recharged by bringing books to Circulation Desk
Renewals and recharges may be made 4 days prior to due date

DUE AS STAMPED BELOW

REC. CIR. MAR 2 '87

FEB 15 1986

REC CIRC MAR 11 1986

UNIVERSITY OF CALIFORNIA,
FORM NO. DD6, 60m, 12/80   BERKELEY, CA 94720

(510) 642-6753
- 1-year loans may be recharged by bringing books to NRLF
- Renewals and recharges may be made 4 days prior to due date.